LASER RANGING TECHNIC

Narain Mansharamani

BookSurge Publishing
7290-B Investment Dr
N. Charleston, SC 29418

Published by
BookSurge Publishing
7290-B Investment Dr
N. Charleston, SC 29418

ISBN: 1-4196-9873-7
EAN13: 9781419698736

Citation
Narain Mansharamani, *Laser Ranging Techniques* is revised edition of monograph *Elements of Laser Ranging Techniques (2001)* ISBN – 81-211-0274-X

Cover Photo
Nd:YAG Ranger cum Designator

PREFACE

The object in writing this monograph is for the scientist, engineers, student and users highlighting the various techniques of laser ranging, considering technological complexities and possible solution.

Initially laser range finders were developed for finding accurate range of non cooperate military targets, in order to give ballistic correction to guns for first round hit on targets, later on with improved ranging accuracy, ranging of co-operated targets with retro-reflector mounted for survey purpose has been carried out within millimeter order of accuracy. Now with the development of tunable laser sources from ultraviolet to mid infrared, ranging techniques have been used for air pollution monitoring ozone layer range and concentration studies, meteorology for remote, temperature, pressure, air speed and cloud study/humidity measurement. For space use, it is used as altimeter for topographical mapping of earth and planets, in oceanography for depth measurement of shallow water and detection of submerged objects. It is also used for earth drift measurement and early prediction of earthquake. It is also used as remote anemometer for smooth landing of aircraft under sever air turbulence. In range gated imaging, vehicles can be moved safely under poor visible condition. In sports, it is used for accurate measurement of golf ball pit and in civil construction as measuring tapes and bridge layout. In industry ranging is used for accurate control of movement of cranes. Keeping in view various application of ranging various techniques used are covered in this monograph.

This monograph is written in twelve chapters to cover most of ranging techniques and its safe use against eye hazards.

Chapter-1: Fundamental of Light and Laser Basics.
Chapter-2: Propagation of Laser Beam in Atmosphere.
Chapter-3: Type of Laser Sources
Chapter-4: Optical Resonators.
Chapter-5: Laser Pumping
Chapter-6: Q-Switching Techniques
Chapter-7: Electronics
Chapter-8: Optics & Thin Films`
Chapter-9: Low-Level Detection
Chapter-10: Laser Ranging Systems
Chapter-11: Recent Developments in Laser Ranging Techniques
Chapter12: laser Safety.

In revised edition of this monograph, non-linear optics and materials are discussed in chapter-1 for efficient optical generation of light in mid infrared for ranging purpose with development sensitive infrared detectors operating at higher temperature where thermo-electic coolers can be used. Resent development of laser sources using diode pumping and sources developed using Cr^{+4}:YAG as passive Q-Switched element. Laser sources using efficient crystalline host are given and miniature diode pump source Nd:KWG is described for compact laser range finder. Microprocessor based range-measuring unit with first and last echo logic is described.

From time to time new technologies are coming up, therefore for latest knowledge, it is suggested these publications are good source for laser technologies details, Optics letter, Applied Optics (USA), IEEE Journal of Quantum Electronics, Laser Focus World, Photonics Spectra, Applied Physics Letters, some SPIE proceedings and IEEE J of Selected Topics in Quantum Electronics.

I am grateful to various army officers and staff, especially Maj. General D.K.Sen, Commandant, ACC&S, Ahmandnagar for their help in carrying out field trials of laser range finders developed under my guidance.

I am grateful to various Directors Generals of Defense Research & Development Organization (DRDO) and Directors, Instruments Research & Development Establishment (IRDE), Dehradun, for their interest in laser development work and allotment of liberal funds for laser range finder development. Especially Dr.A.P.J. Abdul Kalam, Director General DRDO (1992-97) for his interest in high repetition rate laser range finder for various service roles and Mr. S.S. Sundram present Director IRDE for giving permission to high light recent development work carried out.

I am grateful to engineers of various production agencies, especially, Mr.A.R.Vaidya, Additional General Manager, M/S Bharat Electronics, Pune for translating our development work in production of these range finders.

I am grateful to staff and officers of IRDE for help in development activities of laser range finders and for writing this monograph, especially Mr. Ikbal Singh, Scientist F for optical design, Mr. G.P. Dimiri, Scientist F for his contribution on optics technology, Dr. P.Kant, Scientist for thin film coating procedure and Mr.S.P.Gaba, Scientist F, Dr S.N. Vasan Scientist E, and Mr. Neeraj Bhargava (Laser Division) for their recent development work for electronics, lasers and mechanical design of laser division, especially for revised edition of this monograph.

In the end, I am grateful to my son Mr. Rohit, who has presented laptop, printer and scanner to me on father's day. This has made my writing of revised edition easier. This book could not have been written without the encouragement, patience and support of my wife Sarla.

Dehradun, India
June 2008

Narain Mansharamani
E-mail: narainm@hotmail.com

CONTENTS

CHAPTER-1

FUNDAMENTALS of LIGHT & LASER BASICS

1.1 – EARLIER HISTORY ON THEORY OF LIGHT

In 1666, Sir Isaac Newton, after his famous experiment on prismatic decomposition of white light into its component colors, elaborated, what is known as corpuscular theory of light. According to corpuscular theory, "Light is regarded as flight of material particles emitted by source, the sensation of light being produced by their mechanical action upon the retina". The rectilinear propagation followed once from the second law of motion. Reflection and refraction and color in thin film he explained by putting rays of light into Fits i.e. corpuscles being supposed to arrive at the surface in different phase i.e. Fits of easy reflection and easy transmission. The small bodies which by their attractive powers or some other force, stir up vibrations in which they act, which vibration being swifter than the rays, overtakes successfully, and agitate them so as by turn to increase and decrease their velocities thereby put them into these Fits [1,2].

In 1676, Ole Roemer demonstrated that light requires a finite time for its propagation, traveling across space with a velocity, which he estimated at 192,000 miles per second. Now the impact of corpuscles moving at such high speed might well expected to exert a pressure.

Although, optical pressure could not be detected at that time, due to corpuscular striking surface, Christian Huygens 1678 gave wave theory suggested that light results from the molecular vibration in the luminous material. Further, that vibrations were transmitted through 'Ether' as wavelike movements (like ripples in water). Huygens concluded that the result of these transmissions acted on the retina, stimulating the optic nerves to production vision. Huygen, considering wave theory, explained reflection, refraction, and double refraction in unaxial crystal by wave theory. But Newton clung to idea of corpuscles, presumably, as he could not visualize strong medium, which could be defined, if wave theory was given. Later discovery of polarization of light was blow to Huygen"s longitudinal wave theory, which later on Augustin Jean Frensel in 1817 gave concept for transverse wave and explained interference of light. He had to define elastic solid medium Ether, which can sustain waves at such high speed with no obstruction to material bodies, but millions of times rigid than steel. Although no evidence of longitudinal disturbance was observed in Ether, since transverse one in elastic Ether medium must be associated with longitudinal disturbances.

In 1821, H.C. Osterd observed magnetic effect of current and Michael Faraday conception of electric and magnetic force and their interrelation expressed in terms of lines of force, were founded and well known. Faradays Laws of electromagnetic induction were established in 1831. From this James Clerk Maxwell developed the equation that underlines modern theory of electromagnetic waves.

In 1860, Maxwell showed that the propagation of light could be regarded as an electromagnetic phenomenon; the wave consists of an advance of coupled electric and magnetic forces. If an electric field is varied periodically, a periodical magnetic field is obtained, which in its turn generates a varying electric field and so the disturbances is passed on in the form of a wave, electric forces generating magnetic and magnetic generating electric. Maxwell's theory predicated the speed at which these electromagnetic waves would travel from measurements of the magnetic field of electric current, the velocity of propagation being the ratio of the electromagnetic to the electrostatic unit. This ratio, determined, from electric and magnetic measurements, turned out to be the velocity of light, and indicated that light was essentially an electromagnetic phenomenon. Hertz discovered the wave predicated by Maxwell in 1892, which produced electric oscillation in a pair of conductor between which sparks were passed. The conductor radiates waves that could be detected by another pair of similar conductor, which oscillates in resonance with the first pair, causing the passage of minute's sparks between these conductors. Maxwell theory, like the electric solid theory requires Ether, but not a Mechanical in which material displacement took place, but rather an electromagnetic one, in which displacement current and magnetic field could occur. The periodic disturbances, which are suppose to constitute these waves, were called displaced current by Maxwell and these displaced currents can occur in free Ether, or in a dielectric i.e. in non-conductor of electricity. Maxwell theory told as nothing about the nature of this electric displacement, so that in one sense, about the real nature of luminous disturbances were much vaguer, than the wave fifty year earlier, when the elastic solid theory was generally accepted. For in the motion of a solid, we are dealing with perfectly definite physical process. As Schusler remark in the preface of his work on Optics "So long as the character of the displacement, which constitute the waves, remain undefined, we cannot pretend to have established a theory of light" [2].

1.2 – LORENTZ THEORY OF LIGHT SOURCES

Maxwell theory having identified light an electromagnetic disturbance, H.A. Lorentz advanced this idea that the atoms and molecules contain electrons, but capable of vibration under the influence of a restoring force, when displaced. The bound electrons execute damped vibration under the influence of a restoring force, when displaced. The bound electrons, executing damped vibration were supposed to be the source of the luminous disturbances, or it might act as an absorber of radiation energy, when its natural frequency of vibration was in agreement with that of the radiation. This theory accounted for many of the newly discovered effects of magnetism on light, as well as the older phenomena of refraction, dispersion etc.

1.3 – BLACK BODY

Kirchhoff established that at a given temperature, the ratio between the emission ε and absorption powers A for a given wavelength is same for all bodies.

$$\varepsilon / A = \text{constant} \qquad \qquad \dots \dots [1]$$

For all bodies at the same temperature, the more powerfully a body absorb, more powerfully, it will emit when heated i.e. black body will give more light when heated. A black body will absorbs all electromagnetic radiation that falls on it, such an object if heated would be perfect radiator, producing radiation which are called black body radiation. Example of black body in laboratory is a container with a hole in which radiation is trapped. If such a container is heated, radiations bounce around inside the cavity and eventually escape from the hole as black body radiation. The important point about black body radiation is that its property depends only on its temperature. The curve representing the spectrum of radiation is shaped like smooth hill as shown in figure-1, with a peak at a frequency, which depends on its temperature. The hotter the radiator, higher the frequency at which radiation occurs.

Emission S'

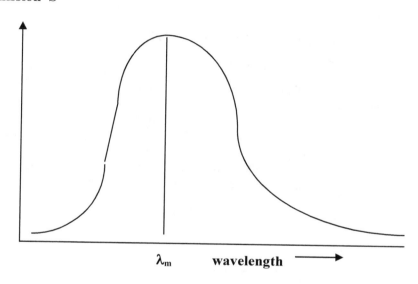

λ_m **wavelength**

Figure-1: Irradiative Spectrum of Black Body at Temperature

Stefan from observations made by others on the intensity of the total radiation emitted from bodies at different temperature established law which states the complete emission S' of a black body is proportional to the fourth power of the absolute temperature T

$$S' = \sigma T^4$$

......[2]

In which σ is a constant?

Wein further observed that as the temperature of a black body increases, the distribution of spectrum of radiation (peak λ_m) shift towards lower wavelength and expressed it by relation known as Wein's Displacement Law

$$\lambda_m T = \text{constant}$$

......[3]

Sun follows the black body curve with a peak correspondence to a temperature under 6000^0 K.

In order to explain, Wein considered black body as cubic box with dimension L', he calculated total number of modes per unit volume between $v = 0$ and v. The mode density per unit frequency $\rho[v]$ is given by

$$\rho[v] = \frac{1}{V} \cdot \frac{dN[v]}{dv} = \frac{8\pi v^2}{c^3} \qquad \dots\dots\dots[4]$$

The spectral distribution has peak at a particular temperature and its intensity decreases on both sides, which could not be explained by mode density.

1.4 – QUANTUM THEORY OF LIGHT

Since the spectral distribution has peak at a particular temperature and its intensity decreases on both sides, which is not explained by mode density. Max Plank introduced concept of discrete modes in 1900 known as Plank's hypothesis: "Energy in electromagnetic wave is in the form of discrete packets that he called quanta's. He further concluded that the size of the quanta for any partial form of electromagnetic ratio is direct proportional to its frequency. The ratio between quanta size and frequency is symbolized as h, one of the fundamental quantities of universe". This suggested that light had a corpuscular nature and average energy per mode E, is given by

$$E = \frac{\sum\limits_{N=0}^{\infty} Nhv \exp[-Nhv/kT]}{\sum\limits_{N=0}^{\infty} \exp[-Nhv/kT]} = \frac{hv}{\operatorname{Exp}[hv/kT] - 1} \qquad \dots\dots[5]$$

so that mode density per unit frequency $\rho[v]$ is given by

$$\rho[v] = \frac{8\pi hv^3}{c^3} \cdot \frac{1}{\exp[hv/kT] - 1} \qquad \dots\dots\dots[6]$$

and average number of quanta [photon] per mode N so that

$$N = \frac{E}{hv} \cdot \frac{1}{\exp[hv/kT] - 1} \qquad \dots\dots[7]$$

This hypothesis gave birth to Quantum Theory, as distribution of radiation density in a cavity in thermal equilibrium at absolute temperature T.

$$\rho[\nu]\,d\nu = \frac{8\pi h\nu^3}{c^3} \cdot \frac{d\nu}{\exp[h\nu/kT] - 1} \qquad \ldots\ldots\ldots[8]$$

Agreed with the measured spectral intensity with frequency as measured at temperature T.

As early as 1754, an attempt was made by D. C. Mairan and Du Fray to detect the **pressure of light** to verify corpuscular theory, but pressure could not be detected. The pressure was first observed by Lebedore in 1900 and later measured by Poynting in 1904 and result came in close agreement as predicted by Maxwell Theory i.e. pressure exerted on unit area is equal to energy contained in unit volume of the vibrating medium. Even deflection of ray of star was observed due to gravitational field of sun observed during total solar eclipse, evidence in favor of Newton's corpuscular theory of light.

In 1905, Albert Einstein could able to explain **photoelectric effect** on concept that electron absorb whole photon i.e. inelastic collision between electron in metal and quantum of light. Later **Compton effect** i.e. elastic collision between quanta of x-ray and electron was explained by Compton i.e. shift of wavelength of x-ray and change in electron kinetic energy.

In 1913, Niel Bhor gave the **structure of atom** after estimating size of nucleus by α–particle scattering experiment by Rutherford in 1912 and emission of hydrogen spectra from Plank's Law.

1.5 – BOHR's HYPOTHESIS

[a]. For each atom, there are series of orbits in which the electrons rotate, but no radiation is emitted. These are called stationary states i.e. angular momentum of the electron in these states is an integral multiple of $h / 2\pi$, where h is Plank's constant.

In the hydrogen atom, we have one electron and one proton. Let the charge on the electron be -e, and radius, the centrifugal force F is given by

$$F = \frac{m_e v^2}{r} \qquad \ldots\ldots\ldots\ldots[9]$$

This must be equal to Coulomb force of attraction between the proton and electron given by

$$F' = \frac{e^2}{r^2} \qquad \dots\dots\dots\dots [10]$$

For balancing of the force in stationary orbit

$$\frac{m_e v^2}{r} = \frac{e^2}{r^2} \qquad \dots\dots\dots\dots [11]$$

or $\quad e^2 = m_e v^2 r \qquad \dots\dots\dots [12]$

According to Bohr's assumption, the angular momentum of the electron is always integral multiple of $h / 2\pi$ here

$$m_e v r = \frac{n h}{2\pi} \qquad \dots\dots\dots [13]$$

$$v^2 = \frac{n^2 h^2}{4\pi^2 m_e^2 r^2} \qquad \dots\dots\dots [14]$$

$$r = \frac{n^2 h^2}{4\pi^2 m_e e^2} \qquad \dots\dots\dots [15]$$

for accepted value of $h = 6.626 \times 10^{-34}$ Joules seconds,
$m_e = 8.98 \times 10^{-28}$ gm
$e = 4.774 \times 10^{-10}$ e.s.u.
$n = 1, 2, 3$, etc.
$r_1 = 0.53 \times 10^{-8}$ cm, $r_2 = 2.12 \times 10^{-8}$ cm etc.

b. The second assumption is that when an electron passes from one orbit, of energy E_1, to another orbit of energy E_2, monochromatic radiation of frequency is emitted, or absorbed, equal to amount $h\nu$

$$E_1 - E_2 = h \nu \qquad \dots\dots\dots [16]$$

As far back as 1885, Balmer noticed, later Lyman, Paschen and Brackett found spectral lines of hydrogen atom in infrared, visible and ultraviolet related by simple relation

$$\nu_0 = 109678.8 \left[\frac{1}{n^2} - \frac{1}{m^2} \right] \qquad \dots\dots\dots [17]$$

i.e. for n = 1, m = 2, 3, 4 ...
   for n = 2, m = 3, 4, 5 ...
   for n = 3, m = 4, 5, 6 ...
   for n =4, m = 5, 6, 7 ...

In order to explain, these lines and multiplexes in the spectrum proposed that electron can move in circular and elliptical orbits depending upon value of principal quantum number, i.e. for n = 4, there is one circular orbit with three elliptical orbits, thus for principal quantum number n, there are n - 1, n - 2, ... , l orbital quantum number.

For l = 0,   s-sharp
   l = 1,   p-principal
   l = 2,   d-diffuse
   l = 3,   f-fundamental series

Further, electrons in atom are characterized by principal n, orbital l, magnetic m, and spin s quantum numbers. The eccentricity of orbit increases with orbital quantum number l, i.e. if l = 4, ratio of radius of major to minor axis is 4. The quantum number m describe the orientation of the orbital momentum vector with respect to an external field, it may assume the values - l to + l, that is for fixed value of l, a total of 2l + 1 value of m are possible. The spin quantum number takes value of s as $\pm 1/2$ i.e. in unit of h / 2π. The total angular momentum of most atoms are obtained by first adding vectorically the orbital angular momentum of the individual electrons and combing the spin angular momentum separately. The total angular momentum combine into a vector L, where magnitude is an integer; the spins combine into a vector characterized by S, where magnitude is an integer for an even number of electrons and a half integer for an odd number. The total angular momentum J of the atoms may be obtained by vector addition of L and S. For each fixed value of S there are 2S+1 different possible spin configurations. With S = 0, the states are called singlets, those with S = 1/2 are doublets etc. For an atom with two electrons, S is either 0 or 1; therefore, such an atom will have singlet and triplet states. The collection of states with common values of J, L, and S is called a term. The symbol characterizing the term is of the form $^{2S+1}X_J$, where X stands for S, P, D, F, G, I etc. depending on the value of orbital angular momentum i.e. L= 0, 1, 2, 3, 4, 5 etc. Different elements of the multiplet are distinguished by J, the total angular momentum J, varies in integer steps from L − S to L + S, but when L = 0, then J = 1 is then only possible value, for S = 1, J has three values L - 1, L, L + 1.

The states of electron in an atom is defined by the Pauli Exclusion Principal, which states that no two electrons in atom may not possess completely identical sets of quantum numbers.

1.6 – MATTER WAVES

In 1914, de Broglie put forward the new idea that all moving material particles of whatever nature have wave-properties associated with them. These particles will, of course, include electrons. Newton in his corpuscular theory to explain Newton's rings, have anticipated periodicity. To this extent he may be said to have anticipated the particle-wave-motion theory. Wavelength λ, as the property of wave, and momentum mv as the property of particle, the two are connected by the de Broglie equation

$$\lambda = \frac{h}{mv} \quad\quad\quad\quad\quad ....[18]$$

Since mv is the moment of particle i.e. for electron with accelerating potential of 100 volts is found to have a wavelength of 1.22 A°. The electron microscope is most outstanding evidence that electron are associated with the waves.

1.7 – BOLTZMANN's STATISTICS

When there are large number of atoms in thermal equilibrium at temperature T, population of atoms at energy levels E_1 and E_2, is given by Boltzman distribution as

$$\frac{N_2}{N_1} = \exp - \left[\frac{E_2 - E_1}{kT} \right] \quad\quad\quad ....[19]$$

Equation [19] is valid for non-degenerate energy levels. If g_1 and g_2 are number of energy levels with same energy in E_1 and E_2 energy levels of atoms i. e. the multiplicity of the energy level of atom called degenerate, then distribution of atoms in thermal equilibrium is given by equation [20]

$$\frac{N_2}{N_1} = \frac{g_2}{g_1} \exp - \left[\frac{E_2 - E_1}{kT} \right] \quad\quad\quad ....[20]$$

1.8 EINSTEIN THEORY OF INTERACTION OF RADIATION WITH MATTER – CONCEPT OF STIMULATED RADIATIONS

Einstein in 1917, combining Plank's law and Boltzman's statistics formulated the concept of stimulated emission [7]. Let monochromatic radiation with density $\rho[v_{12}]$ interact with atoms of matter with energy levels E_1 - E_2 such that

$$E_2 - E_1 = hv_{21} \quad\quad\quad ....[21]$$

In order to consider interaction of radiation with matter, Einstein defined the radiative rate coefficients as:

A_{21}: Spontaneous emission coefficient i.e. probability per second that an atom in a state E_2 will make transition to a state at E_1 without any outside influence i.e. no electromagnetic field is required. It is inverse of the spontaneous lifetime of atom in state 2 i.e. τ_s.

B_{12} is absorption coefficient of material i.e. the probability per second an atom will make transition from lower level E_1 to upper level E_2 with absorption of quanta of a radiation is

$$B_{12}\,\rho[\nu_{12}] \qquad\qquad\qquad\qquad\qquad \dots\dots\dots[22]$$

The probability per second an atom will make transition to lower state at E_2 with emission of quantum of energy $h\nu_{21}$ is given by

$$A_{21} + B_{21}\rho[\nu_{21}] \qquad\qquad\qquad\qquad \dots\dots\dots[23]$$

B_{21} he called as stimulated emission coefficient.

Thus at any time the total number of atoms N_{TH}

$$N_{TH} = N_1 + N_2 \qquad\qquad\qquad\qquad\qquad \dots\dots\dots[24]$$

$$N_2 = B_{12}\,\rho[\nu_{12}] \qquad\qquad\qquad\qquad\qquad \dots\dots\dots[25]$$

$$N_1 = A_{21} + B_{21}\rho[\nu_{21}] \qquad\qquad\qquad\qquad \dots\dots\dots[26]$$

According to equations [19], [21], [25] and [26]

$$\frac{B_{12}\rho[\nu_{12}]}{A_{21} + B_{21}\rho[\nu_{21}]} = \exp.\frac{-h\nu_{21}}{kT} \qquad\qquad \dots\dots\dots[27]$$

If $\nu_{12} = \nu_{21} = \nu$

$$\rho[\nu] = \frac{A_{21}}{B_{12}} \cdot \frac{1}{\exp.[h\nu/kT] - B_{21}/B_{12}} \qquad\qquad \dots\dots\dots[28]$$

Comparing equation with Plank's law relation [6] we have

$$B_{21} = B_{12} \qquad\qquad\qquad\qquad\qquad \dots\dots\dots[29]$$

$$\text{and } A_{21} = \frac{8\pi h\nu^3}{c^3}\,B_{12} \qquad\qquad\qquad \dots\dots\dots[30]$$

The relations between A's and B's are known as Einstein relations and the factor

$$\frac{8\pi\nu^2}{c^3} \text{ is mode density } p_n \qquad\qquad\qquad \dots\dots\dots[31]$$

Normally at low temperature more atoms are at ground state E_1. As a consequence of the absorption of radiation, the equilibrium of atoms in ground state will be disturbed. Atoms that become excited due to absorption of radiation may return directly to ground state by spontaneous or stimulated radiation, or they may follow another path and change to a lower level other than the ground level. In this manner they may cascade down on the energy scale, emitting at each step radiation different in frequency from that, which originally lifted them out of the ground state. Because of the energy and frequency relation, the radiation emitted in the cascade process, which is called fluorescence, has lower frequency than the exciting radiation. At any such time, due to absorption of radiation, population of atoms in E_2 State is more than ground state; condition is known as population inversion.

1.9-BOSE-EINSTEIN STATISTICS

S.N.Bose in 1920 introduced distribution of photons [9], which was supported by Einstein and published as Bose-Einstein Statistics. Unlike other particles, photons can occupy the same state at the same time. At low temperature, all the photons will tend to congregate together at the same lowest energy state, forming what is known as Bose-Einstein condensate. The probability that photons N_i will have energy E is given by

$$N_i(E) = \frac{1}{A'e^{E/kT} - 1} \qquad \ldots\ldots[32]$$

For photons $A' = 1$, therefore at low temperature occupancy of photons at low energy state can increase without limit forming what is known as Bose-Einstein condensate.

1.10 – ATOMIC LINE SHAPE FUNCTION

(A change in the frequency of light, or in the pitch of sound, caused by the motion of object emitting light or making the sound is called Doppler effect)

Atomic line shape function g(v) is defined as the probability that a given transition will result in emission (or absorption) of a photon with energy between dv to (v+dv) [6,7].

For normal absorption or emission it follows that

$$g(v)\,dv = 1 \qquad \ldots\ldots\ldots[33]$$

Absorption or emission intensity is plotted against v. The origin of this distribution may be due to spread in the Doppler shift as in the case of low-pressure gas due to velocity of gas molecule at temperature T.

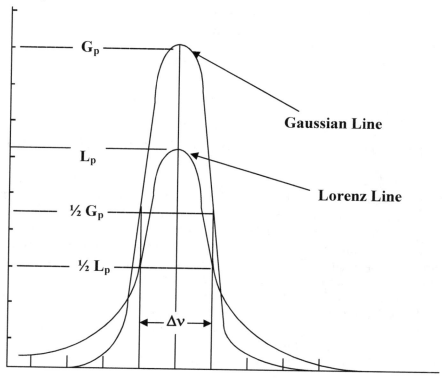

Figure-2: Atomic Line Wave Function

i.e. $\exp[\, -m_a v_x^2/kT\,] \cdot v_x$

Doppler shift $\dfrac{v_x}{c} = \dfrac{v-v_0}{v_0}$ ...... [34]

$g(v)\, dv = \rho(v_0)\, \exp.\, \{\, -\beta(v-v_0)^2/v_0^2\}\, \cdot dv$ ....[35]

$\rho(v_o) = c/v_0\, (m_a\, /\, 2\pi kT)^{1/2}$ ....[36]

$\Delta v = 2v_0/c\, (2kT\, \log2/m_a)^{1/2}$ ....[37]

m_a = mass of atom

In case of solid due to small lattice vibration distribution due to atoms broadening (Lorentzian) line is given by

12

$$g(v) = \frac{\Delta v}{2\pi} \frac{1}{(v-v_0)^2 + (\Delta v/2)^2} \qquad \ldots\ldots\ldots [38]$$

Where Δv is Line-width at half-maximum.

The atomic line shape function plot i.e. intensity versus frequency is shown in figure-2

1.11 - ELEMENTARY CONSIDERATION REGARDING LASER ACTION

It is well established from Plank's law, Boltzman statistics, Einstein interaction of radiation with matter and Bohr emission theory of radiation that atoms can be excited to higher energy state, if it interacts with photon having same energy corresponding to energy levels of atom. Alternately, if the photon of same energy interacts with excited atom, it will come down to lower energy level, emitting additional photon of same state as interacting photon. Therefore, in order to establish laser, C.H.Townes proposed an optical cavity consisting of parallel mirrors with active medium and source to excite laser material as shown in figure-3. Active medium can be excited to have more atoms in excited state, so that photon can be confined between mirrors and can interact with active excited medium. The cavity can be defined by quality factor i.e. energy stored to energy lost per cycle or decay time of photon τ_d in the cavity. If V is volume of cavity, $\rho(v)$ photon density in cavity, then energy stored in cavity is given by

$$E = \rho(v) \, hv \, V \qquad \ldots\ldots\ldots[39]$$

Energy lost per cycle, is given by relation

$$E_{Lost} = \frac{\rho(v) \, hv \, V}{\tau_d} \cdot \frac{1}{2\pi v} \qquad \ldots\ldots[40]$$

$$\text{The Quality factor } Q = \frac{\text{Energy stored}}{\text{Energy lost per cycle}} \qquad \ldots\ldots[41]$$

$$Q = \frac{E}{E_{Lost}} = 2\pi v \tau_d \qquad \ldots\ldots[42]$$

The induced emission depend on the number of atoms n_1 and n_2 in energy state E_1 and E_2 of atom

The induced emission rate in the frequency interval dv around v is given by

$$hv \, g(v) \, (B_{21} N_2 - B_{12} N_1) \, dv \qquad \ldots\ldots [43]$$

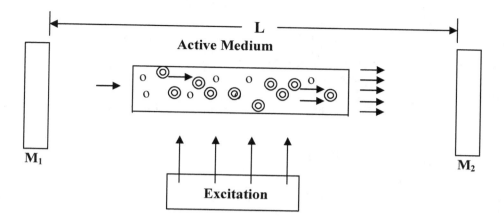

Figure-3: Optical Cavity with Active Medium (Proposed by Townes)

If energy levels E_1 and E_2 have degeneracy factor g_1 and g_2

Then $B_{12} = B_{21} (g_2/g_1)$ ....[44]

The induced emission rate can be written as

$h\nu \, g(\nu) \, \rho(\nu) \, B_{21} \, [\, N_2 - N_1 \, (g_2/g_1) \,] \, d\nu$ ....[45]

If $2\Delta\nu$ is line width at half maximum of broadened emission, then for Lorentzian distribution

$$g(\nu) = \frac{\Delta\nu}{\pi} \left[\frac{1}{(\nu - \nu_0)^2 + (\Delta\nu)^2} \right] \qquad ....[46]$$

at $\nu = \nu_0$, $g(\nu_0) = 1/(\Delta\nu\pi)$ ....[47]

For broadened Lorentzian line rate of induced emission is given by

$$\frac{d\rho(\nu_0)}{dt} = \frac{h\nu_0 \, B_{21}}{\Delta\nu\pi} \, \rho(\nu_0) \left[N_2 - N_1 \left\{ \frac{g_2}{g_1} \right\} \right] \qquad ....[48]$$

If τ_d is cavity-ringing time, rate equation for induced emission can be given as

$$\frac{d\rho(\nu_0)}{dt} = \frac{h\nu_0 \, B_{21}}{\Delta\nu \, \pi} \left[N_2 - N_1 \left\{ \frac{g_2}{g_1} \right\} \right] \rho(\nu_0) - \left[\frac{\rho(\nu_0)}{\tau_d} \right] \qquad ....[49]$$

On integration we obtain growth of photon density of induced emission as

$$\rho(v_0) = \rho_0 \exp. \left[\frac{h v_0 B_{21}}{\pi \Delta v} \left\{ N_2 - N_1 \left(\frac{g_2}{g_1} \right) \right\} - \frac{1}{\tau_d} \right] t \qquad ....[50]$$

For growing photon density i.e. threshold condition of induced or stimulated emission

$$\frac{h v_0 B_{21}}{\Delta v \pi} \left(N_2 - N_1 \left\{ \frac{g_2}{g_1} \right\} \right) > \frac{1}{\tau_d} \qquad ....[51]$$

$$\text{Or} \quad \left(N_2 - N_1 \left\{ \frac{g_2}{g_1} \right\} \right) > \frac{\Delta v \pi}{h v_0 B_{21} \tau_d} \qquad ....[52]$$

If τ_s is emission lifetime of metastable state i.e. excited state of atoms E_2 then

$$A_{21} = \frac{1}{\tau_s} \quad , \quad B_{21} = \frac{c^3 A_{21}}{8\pi h v_0^3} \qquad ....[53]$$

For stimulated emission number of atoms N_2 in excited state at any time should be such that

$$\left(N_2 - N_1 \left\{ \frac{g_2}{g_1} \right\} \right) > \text{or} = \frac{\tau_s}{\tau_d} \frac{8\pi^2 v_0^2 \Delta v}{c^3} \qquad ......[54]$$

For low threshold Δv should be low or line width should be sharp. Laser basically consists of active material, cavity and pumping source, these are discussed in reference [4, 5, 6, 7] in details.

1.12-ACTIVE MATERIALS

Active material may be solid state, liquid or solid dye, non-linear crystal, gaseous like Carbon dioxide (CO_2), Helium-Neon (He-Ne) gas mixture, junction semiconductor or electrons confined in quantum state of thin semiconductor. Excitation source may be flash lamp, laser diode or laser pump for non-linear crystal, injection current drive in semiconductor, electric discharge or by chemical reaction in gases. These are discussed in details in third, fourth and fifth chapter of laser source, optical resonator and laser pumping.

1.13 - PROPERTIES OF LASER BEAM

Unlike ordinary light, following important properties of laser beam was observed.

1.13.1 Directivity: A laser beam is confined to a narrow cone of angles, from a few tenths of milli radians to a few milli radians.

1.13.2 Radiance: Radiance is the radiant power (W) per unit solid angle (S_r) per unit surface area (m^2) that is emitted from an infinitesimal area of a surface into a cone perpendicular to the surface. Laser beam has 1 to 10 order higher radiance that of most intense incoherent source.

1.13.3 Coherence: Coherence is the orderly relationship of one part of the beam to another part.

[i] Spatial Coherence: If electromagnetic (e.m.) field at two points oscillates with a fixed relation to each other, than the e.m. field at all points is said to be spatially coherent.

[ii] Temporal Coherence: If at any point in space, the electro magnetic field oscillates in stable and predicable manner from one point in time to another point in time, than there is perfect temporal coherence.

1.13.4 Sharp Pulses: The process of fixing the frequency and phase difference is defined as mode locking. Mode locking can produce sharp pulses.

1.13.5 Phase Conjugation: Phase conjugation is the phenomena in which direction of phase and direction of wave is reversed. Phase conjugation in a material can be produced by virtue of non-linear interaction using photo refractive effect or stimulated Brillouin scattering by intense laser beam.

1.14 – INTERACTION OF INTENSE COHERENT OPTICAL RADIATION

1.14.1 Nonlinear Optical Effects

Nonlinear optical effects arise from the interaction of intense coherent optical radiation with matter. These effects are harmonic frequency generation, sum and difference frequency generation, and optical parametric amplification and oscillations; stimulated Raman scattering and stimulated Brillouin scattering; refractive index changes and self focusing; three wave and four wave mixing and phase conjunctions; self induced transparency in absorption medium e.g. optical saturable absorption in organic dyes and in semiconductors: changing reflectivity; two photon absorption (TPA); fluorescence, ionization and dissociation; and optical cooling [10].

These non-linear effects can be explained by considering non linear electric polarization of medium or by quantum theory if radiation is in resonant with atom or molecules of medium, Non-linear electric polarization P arises due to displacement of charges in atoms / molecules of optical medium as a result of intense electric field E of coherent radiation

$$P = \varepsilon_0 [\chi^{(1)} E + \chi^{(2)} E E + \chi^{(3)} E E E + \text{---------}] \qquad \text{......[55]}$$

where ε_0 is permittivity of free space and $\chi^{(1)}$, $\chi^{(2)}$, $\chi^{(3)}$ are called 1st order, 2nd, 3rd susceptibility, These are dielectric coefficients of medium and are tensors.

In case of resonant interaction of radiation with matter, a concept of virtual energy level can be introduced, which represent an intermediate quantum state occupied by the combined system of the photon field and molecules. Based on the concept of virtual energy level or intermediate state, the principal and mechanism can be considerably interpreted, and in many cases, clearly illustrated by an energy level diagram involving the transition via virtual energy level.

2nd order susceptibility of materials in non-centro symmetrical crystals are responsible for 2nd harmonic generation, sum and difference frequency generation, optical parametric amplification and oscillations. 3rd order susceptibility of material is responsible for 3rd harmonic generation, three wave and four wave mixing to generate phase conjugate wave.

Resonant Raman and Brillouin scattering, two photon absorption (TPA) can be interpreted and explained by quantum theory of light.

1.14.2 Parametric Oscillator

When two beams with different frequencies are incident on a non-linear crystal, a polarization wave at difference frequency is generated. If polarization wave travels at the same velocity as freely propagating electromagnetic wave, cumulative growth will take place. Two incident beams are termed as pump and signal waves having frequency ν_p and ν_s then third polarization wave is termed as idler wave with frequency ν_i. Under certain condition, idler wave can mix with the pump beam to produce a polarization wave at the signal frequency, phased such that growth of the signal wave result. The process continues with the signal and idler wave both growing and pumps wave decaying as a function of distance in the crystal. As each pump photon with energy $h\nu_p$ is generating a photon at the signal $h\nu_s$ and idler frequency $h\nu_i$, energy conversation requires

$$\nu_p = \nu_s + \nu_i \qquad \text{......[56]}$$

To achieve parametric amplification, polarizing and electromagnetic wave should propagate at same velocity.

For momentum matching condition

$$k_p = k_s + k_i \qquad \text{.......[57]}$$

$$\frac{n_p}{\lambda_p} = \frac{n_s}{\lambda_s} + \frac{n_i}{\lambda_i}$$

.......[58]

Where n_p, n_s and n_i are refractive indices at the pump, signal and idler frequencies respectively. If refractive indices are varied, the signal and the idler frequencies will vary. For tuning, it is possible to make use of the birefringence of nonlinear crystals.

The generation of **harmonics** is a special case of optical mixing in nonlinear materials.

SUMMARY/DEFINATIONS

1. **Corpuscular Theory of light:** Light is regarded as flight of material particles emitted by source.

2. **Electromagnetic Theory of Light**: The propagation of light can be regarded as an electromagnetic phenomenon; the wave consists of an advance of coupled electric and magnetic forces. If an electric Field is varied periodically, magnetic field is obtained, which in turns generate a varying electric field and so the disturbances are passed in the form of wave, electric force generating magnetic and magnetic generating electric.

3. **Black Body:** A hypothetical object, which absorbs all electromagnetic radiation that falls on it. Such an object, if heated, would be a perfect radiator, producing black body radiation.

4. **Black Body Radiation:** The radiation emitted by hot black body. The best example of a black body in the laboratory is a container with a small hole in it, into which radiation shines and trapped. If such a container is heated radiation bounces around inside the cavity and eventually escape from the whole as black body radiation.

5. **Kirchoff Radiation Law:** Kirchoff radiation law states that at fixed temperature, the ratio between emission and absorption power of surface for a given wavelength is same.

6. **Stefan Law:** Stefan law states that complete emission from black body is proportional to fourth power of absolute temperature.

7. **Wein Displacement Law**: Wein displacement law states peak wavelength in the distribution of spectrum of radiation is inversely proportional to absolute temperature of black body.

8. **Plank's Hypothesis**: Energy in an electromagnetic wave is in the form of discrete packets that he called quanta's. The size of quanta is directly proportional to its frequency. The ratio between quanta size and frequency is symbolized as h, one of the fundamental quantities of universe.

9. **Neil Bohr's Emission and Absorption Theory**: Electrons in an atom rotates in a stationary orbit, no radiation is emitted i.e. in these orbits angular moment of electron is integral multiple of $h/2\pi$, electron can absorb photon of frequency ν, excited to energy state E_2 from E_1, given by relation $E_2 - E_1 = h\nu$ or it can emit quanta of energy when making transition from state E_2 to E_1.

10. **Degeneracy:** When the electrons defined by different quantum number have same energy state, it is known as degeneracy, it is also known as multiplicity of energy state of atom.

11. **Pauli's Exclusive Principal:** An orbiting electron in an atom is defined by four quantum numbers, principal quantum number n, orbital quantum number l, magnetic quantum number m_l and spin quantum number s. According to Pauli's exclusive principal no two electron in an atom have same set of four quantum numbers.

12. **Selection Rule**: For orbital quantum number, transitions of electron are possible only between terms n, l and n, l+1, no restriction being imposed on the number n and m_l by the selection rule.

13. **Excited State:** The state of an atom in which it has greater energy than in the ground state.

14. **Absorption Spectra**: It corresponds to transition of atoms from ground to the excited state.

15. **Emission Spectra**: Emission spectra correspond to transition from excited state characterized by a long lifetime known as **meta stable state** to a lower energy level known as **terminal level** of the atom.

16. **Birefringence:** Difference in velocity of light with direction of polarization in a nonlinear crystal is known as birefringence.

17. **Boltzman Statistics:** Applicable to ions, molecules, and atoms. Similar atoms in thermal equilibrium at temperature T, the relative populations N_1 and N_2 at any time in the energy level is given by

$$\frac{N_2}{N_1} = \exp{-\left(\frac{E_2 - E_1}{kT}\right)}$$

18. **Bose-Einstein Statistics:** The probability that photons will have energy E is given by

$$N_i(E) = \frac{1}{A'e^{E/kT} - 1}$$

For photons A' =1, therefore at low temperature occupancy of photons at low energy

19. **Population Inversion:** When more atoms or molecules are in excited state as compared to ground state under influence of exciting energy source, the material is in state of population inversion.

20. **Mode Locking:** The process of fixing the frequency separation and phase difference is defined as mode locking.

21. **Optical Pumping:** When the active material is subjected to extreme intense light source, which causes more transition of atoms or molecules to excited state, at the same time transition can occur to lower level, this excitation by intense light source is known as optical pumping.

22. **Quality Factor:** When an electromagnetic wave is confined in a close optical cavity with active laser material, then the quality factor Q is defined as ratio of stored energy to energy lost per cycle in a cavity.

23. **Fluorescence:** In fluorescence emission from excited state to ground state takes place in cascade process.

24. **Compton Effect:** A change in frequency / energy of photon of monochromatic electromagnetic wave (x-ray) and kinetic energy of free electron due to elastic collision takes place in accordance with conservation of momentum is Compton effect.

25. **Photoelectric Effect:** The increase in kinetic energy of free electron in metal or semiconductor due to absorption of complete energy of photon is known as photoelectric effect.

26. **Coherence:** Coherence is the orderly relationship of one part of the beam to another part.

27. **Spatial Coherence:** In electromagnetic field at two points oscillates with a fixed phase relationship to each other, then the electromagnetic field at all points in the beam are coherent, then the beam is said to be spatially coherent.

28. **Temporal Coherence**: If at any point in space the electromagnetic field oscillates in stable and predicable manner from one point in beam to another point in time, then there is perfect temporal coherence.

29. **Conjugation Phase**: Phenomena in which phase and direction of wave are reversed.

30. **Doppler Effect**: Change in frequency of light or sound due to relative motion between source and observer.

31. **Stimulated Emission**: Emission of photon from excited state of atom or molecule under the influence of radiation of same energy is known as stimulated emission.

32. **Spontaneous Emission**: Emission of radiation from excited atom or molecule without any influence of external electromagnetic field.

33. **Meta Stable State:** In this state excited atoms or molecules have long residential time and spontaneous or stimulation emission takes place from this place to a lower energy level known as terminal level.

34. **Semiconductor**: A semiconductor is a metal with electric conductivity more than good insulator, yet less than metal.

35. **Hole**: A hole is a vacancy of electron in a tetravalent semiconductor.

36. **n-type Semiconductor:** n-type semiconductor has more pentavalent atoms like As, Sb or P.

37. **p-type Semiconductor:** p-type Semiconductor has more trivalent atoms like Al, B, Ga or In.

38. **Band Gap:** Energy difference between free and bound electron in a semiconductor.

39. **Fermi-Level:** Fermi level is energy level in which half of free electron remain in that state.

40. **Brillouin Scattering:** Brillouin scattering involves scattering of light by acoustic phonons, which results from optically induced electrostriction, scatter light is down shifted in frequency.

41. **Raman Scattering:** Raman Scattering involves scattering of light from gas molecules, scatter light is down shifted in frequency that depends on characteristic of gases.

42. **Polarization:** The electromagnetic wave is called polarized wave, when its electric field has fixed orientation.

43. Interference: Interference is a phenomenon in which two or more wave of same frequency overlap in the same medium. If two waves overlap in the same phase, interference is constructive. In destructive interference, waves are completely out of phase.

44. Diffraction: Diffraction is spreading of light around the edges of an obstacle. Diffraction gratings are arrays of identical and equally spaced slits. Diffraction gratings like prism split white light into colors.

45. Frequency: Frequency of a wave is its rate of oscillation and is measured in Hertz.

46. Wavelength: A wave consists of successive trough and crest and wavelength is distance between two adjacent crests.

47. Dipoles: A dipole is created due to displacement of orbiting electrons from nucleus of atom or molecule due to electric field of electromagnetic wave.

48. Photon: A photon is a elementary particle and is also known as quanta of light or boson with spin 1. It has zero rest mass and travel with velocity of light c in free space and has energy $h\nu$ and momentum $h\nu/c$. Like other particle it has dual nature of wave and particle.

49. Refractive Index: Refractive index of a dielectric material is ratio of velocity of light in vacuum to velocity of light in that dielectric medium. Velocity of light in dielectric varies with frequency of light, which is known as dispersion.

50. Susceptibility: A susceptibility of material is ratio of total dipole moment of molecules in unit volume to electric field strength of electromagnetic wave creating dipoles in that material.

51. Wave number: Wave number is reciprocal of the wavelength.

52. Group Velocity is the velocity with which envelope due to superposition of two harmonic waves with different frequency travel in the same direction.

53. Radiant Energy: Radiant energy is the energy of photons or of electromagnetic wave emitted by a body or medium.

54. Radiant Flux is the average energy carried by electromagnetic wave in unit time through an arbitrary surface.

55. Luminous Flux: Luminous flux is radiant flux at light frequency where human eye is most sensitive i.e. 555 nm.

56. Irradiance: Irradiance is equal to the ratio between radiant flux and the area of the uniformly irradiated surface.

57. Phase Matching: In phase matching, index at the fundamental frequency is set equal to value of index of second harmonic by suitable choice of the polarizations in birefringent materials.

58. Self-traping: A nonlinear effect where an increase in intensity creates a change in index of refraction and the focusing generated by this change compensate for diffraction.

59. Nonlinear Optics: A field of optics that exploits the nonlinear relation between the polarizability and the electric field.

60. Stimulated Raman Scattering: Raman Scattering is a non-linear process, where sum and difference frequencies are generated between a pump source and characteristic Raman absorptions in a material. Stimulated Raman scattering differs from spontaneous Raman scattering both in the efficiency of conversion and in the emission pattern. Spontaneous Raman radiation is emitted nearly isotropically. While stimulated Raman radiation is emitted in a narrow cone aligned with the pump beam.

61. Self-Focusing: Self-focusing is a non-linear process where the light intensity causes an increase in the index of refraction. This increase in the index of refraction causes a lensing effect where the beam is focused more tightly. The more tightly focused beam has a higher intensity and thus causes increased focusing.

62. Polarizing angle: For polarized light with polarization normal to plane of reflection, refection is zero at certain angle known as Brewster angle θ and is given by $\mu = \tan \theta$.

63. Light Pressure: Light pressure is mechanical pressure exerted on unit area is equal to energy contained in unit volume of the vibrating medium.
$$p = I_r (1 + \rho) / c$$

64. The Poynting Vector: Poynting theorm states that the time rate of flow of electromagnetic energy per unit area is given by the vector S, called the Poynting vector is defined as the cross product of the electric and magnetic fields,

$$P = E' \times H.$$

Appendix A

Unit of wavelength for Optical Region are

UNIT	ABBREVIATION	EQUIVALENT
micron	μ	10^{-6} m
nanometer	nm	10^{-9} m
angstrom	A^0	10^{-10} m

The Electromagnetic Spectrum

Type of Radiation	Frequency	Wavelength	Quantum Energy
"Wave" region			
radio waves	10^9 Hz and less	300 mm and longer	4×10^{-6} eV or less
microwaves	10^9 to 10^{12} Hz	300 mm to 0.3 mm	4×10^{-6} to 4×10^{-3} eV
"Optical" region			
infrared	10^{12} to 4.3×10^{14}Hz	300 to 0.7 μ	0.004 το 1.7 eV
Visible	4.3×10^{14} to 5.7×10^{14} Hz	0.7 to 0.4 μ	1.7 to 2.3 eV
Ultraviolet	5.7×10^{14} to 10^{16} Hz	0.4 to 0.03 μ	2.3 to 40 eV
"Ray" Region			
x-rays	10^{16} to 10^{19} Hz	300 to 0.3 A^0	40 to 4×10^4 eV
gamma rays	10^{19} Hz and above	0.3 A^0 and shorter	4×10^4 eV and above

Appendix B

Permeability of the vacuum μ_0 is $4\pi \times 10^{-7}$ henries per meter

Permittivity ε_0 of vacuum is $(1/36\pi) \times 10^{-9}$ farads per meter

Velocity of electromagnetic waves c in the vacuum is $(\mu_0\varepsilon_0)^{-1/2} = 3 \times 10^8$ per second.

c = 299,792,456.2 $\pm$ 1.1 m/s [11]

h = 6.626×10^{-34} Joules second (Planck constant)

e = 1.602×10^{-19} Amperes second (Electron charge)

1 eV = 1.602×10^{-19} Joules

k = 1.381×10^{-23} Joules per Kelvin (Boltzmann Constant)

Avogadro Constant, N_A = 6.02204×10^{23} mol^{-1}

Mass of electron m_e = 9.10953×10^{-31} Kg

Mass of Proton m_p = 1.67265×10^{-27} Kg

LIST of SYMBOLS USED

ε = Emission power of surface

A   = Absorption power of surface

σ = Constant of emission

S'   = Complete emission of black body

λ = Wavelength of light

$\lambda_p, \lambda_s, \lambda_i$ = Pump, signal, idler wavelengths

λ_m = Peak radiation wavelength of black body

$p(v)$ = Mode density per unit frequency

T   = Absolute temperature

c   = Velocity of light

v = Frequency of light

V   = Volume of cavity

k   = Boltzman constant

h   = Plank's universal constant

N   = Average number of quanta's (photons) per mode

e   = Charge on electron

m_e = Mass of electron

m_p = Mass of proton

r   = radius of electron orbit

r_1, r_2 = Radius of electron orbit with principal quantum number 1, 2

E_1 = Lower energy level of atom

E_2 = Higher energy level of atom

m_a = Mass of atom

n   = Principal quantum number

m_l = Magnetic quantum number

l   = Orbital quantum number

L   = Total orbital quantum number

J   = Total quantum number

s   = Spin quantum number

S   = Total spin quantum number.

E   = Average energy per mode

L'   = Cavity length

Q   = Quality factor of cavity

$g(v)$ = Atomic line shape function

A_{21} = Spontaneous emission coefficient of atoms in higher energy state

τ_s = Lifetime of atoms in metastable state i.e. in higher energy state

τ_d = Cavity ringing time of photon

B_{12} = Absorption coefficient of material

B_{21} = Stimulated emission coefficient of atoms in higher energy state

g_1 = degeneracy of energy level E_1

g_2 = degeneracy of energy level E_2

N_1 = Number of atoms in energy level E_1

N_2 = Number of atoms in energy level E_2

ε_0 = Permittivity of free space

μ_0 = Permeability of free space
μ = Refractive index
χ = susceptibility
$\chi^{(1)}$, $\chi^{(2)}$, $\chi^{(3)}$ = 1st, 2nd, 3rd order susceptibility
n_p, n_s, n_i = refractive indices at the pump, signal and idler frequencies
λ_p, λ_s, λ_i = Pump, signal, idler wavelengths
P = Pointing vector of electromagnetic wave
I_r = Irradiance of light
ρ = Reflection coefficient of light
p = Light pressure
E' = Transverse electric field of electromagnetic wave
H = Transverse magnetic field of electromagnetic wave

REFERENCES

1. Sir Isacc Newton, "A Treatise of the Reflection, Refraction, Inflection and Color of Light, Forwarded by Albert Einstein, Dover Publication Inc. [Based on 4th Edition 1730].

2. Wood,R.W., Physical Optics, The Mackmillian Company, New York, 1956.

3. Maiman,T.H. "Stimulated Optical Radiation in Ruby Masers", Nature, Vol. 187, August 1960, pp. 493-494.

4. Yariv, A. , and J. P. Gordon, " The Laser, " Proc. IEEE, Vol.51, No.1, January 1963, pp. 4-29.

5. Steele, L. Earl, "Optical Laser in Electronics", John Wiley and Sons Inc., New York, (1968).

6. Encyclopedia Britannica, Vol. 14, pp. 490, Publ. Encyclopedia Britannica, Inc., London, (1980).

7. Lengyel, A. , Bela, "Lasers," Wiley-Interscience, New York (1966).

8. Koechner, Walter, "Solid State Laser Engineering," Springer Series in Optical Sciences, Springer-Verlag, Berlin (1992).

9. John Gribbin, 'Q is for Quantum", An Encyclopedia of Particle Physics, Publisher The Free Press, A Division of Simon & Schuster Inc., 1230 Avenue of the Americans, New York, Great Britain (1998).

10. Guang,S.H., and Lice,S.H., "Physics of Nonlinear Optics", Publisher-World Scientific, Singapore, NJ, London, Hong Kong (1999).

11. Fowlers, Grant R., Introduction to Modern Optics, second edition, Dover Publication Inc., New York, 11501 (1975).

12. Besancon,R.M., (Editor), "Velocity of Light", The Encyclopodedia of Physics. New York: Reinhold, 1966.

13. Kuhn, K.J., "Laser Engineering", ISBN 0-02-366921-7, Prentice-Hall, Inc, NJ (1998).

CHAPTER - 2

PROPAGATION of LASER BEAM IN ATMOSPHERE

2.1 - INTRODUCTION

When highly collimated, monochromatic laser beam travel towards target, it travels to and fro in atmosphere, during its travel it encounter reduction in energy, deviation of its path, increase in beam size i.e. reduction of its collimation or its frequency change due to following effects

1. Resonant absorption by atmospheric gases, pollutants, suspended particles aerosols.
2. Scattering from molecules of atmospheric gases, aerosols, and fog or water droplets.
3. Frequency change due to fluorescence, Doppler effect or Raman scattering from trace impurities.
4. Path deviation of beam due to temperature inversion, air turbulence or random refraction from precipitation and diffraction from aerosol results in reduction of collimation.
5. Intensity of echo received depends on target reflectivity.

Therefore, knowledge of beam propagation is essential in deciding the laser source frequency for a particular environment. Lot of work is going on these days to measure degradation of laser beam parameters at different frequencies for remote pollution control [4], remote aenometery and meteorology. It is difficult to give all atmospheric beam effects in this chapter. Here, only atmospheric beam effects are given which affects range-measuring capability of laser range finders. However some references are given for measurement of atmospheric attenuation, collimation of beam and precipitation.

In this chapter, brief description of atmosphere composition and structure, various types of particles, and sources of impurities, formation of fog, haze, clouds and rain is given. Absorption spectra of atmospheric gases, scattering from molecules and aerosol, attenuation of light in visible and infrared unto 10 micron of light wavelength for various atmospheric conditions of precipitation i.e. fog, clouds and rain mainly responsible for laser propagation are also given.

2.2 - ATMOSPHERIC COMPOSITION & STRUCTURE

Atmosphere / air consists of mainly nitrogen – N_2: 78.084%, Oxygen – O_2: 20.946%, Carbon dioxide – CO_2: 0.0314%, Argon – Ar: 0.934%, Neon – Ne: 18 ppm, Helium – He: 5.2 ppm, Krypton – Kr: 1.1 ppm, Xenon – Xe: 0. 09 ppm, Hydrogen – H_2: 0.5 ppm and Methane – CH_4: 2 ppm. These gases are in fixed proportion, while carbon monoxide – CO: 1.1 ppm, Nitrous oxide – N_2O: 0.5 ppm, Ozone O_3: 0 to 0.07 ppm at ground and 1-3 ppm at 20 to 30 km height, Water vapor – H_2O: 0 –2%, Sulfur dioxide – SO_2, Nitrogen dioxide – NO_2, Ammonia – NH_3, Nitric oxide – NO, Hydrogen sulfide – H_2S and Nitric acid vapor – HNO_3 in minute variable traces depending upon terrain and altitude. The average density of air at sea level is 1.29 kg/m^3. All gases, despite differences in molecular weight have constant volume ratio unto 60,000 meters of altitude because of through mixing by winds and convention currents. Hydrogen and Helium concentration is more in upper atmosphere, while ozone concentration is high at 20 to 30 km height due to its formation from oxygen due to its dissociation by UV radiation. Ammonia is produced in the atmosphere by the decay of organic matter, while hydrogen sulfide and sulfur dioxide are injected into air by volcanoes and hot springs. Sulfur dioxide is also produced in relatively large quantity by the burning of fossil fuel. The oxide of nitrogen i.e. nitrogen dioxide, nitric oxide, nitrous oxide are produced by bacterial activities in soil, by combination of oxygen and nitrogen as a result of lightening, internal combustion engine and in jet engines due to electric ignition. As three – fourth surface of earth is covered with water, water vapor are more over oceans and or seacoast and less over arid or dessert areas. Water exists in vapor, liquid and solid phase in atmosphere. Ice formation is more in upper atmosphere and over arctic regions. Mist and fog formation is due to sudden drop in temperature, while droplets in atmosphere are formed due to condensation of water over hygroscope aerosol in atmosphere. Atmosphere is prevented from escaping due to earth's gravitational pull. Structure of atmosphere is shown in figure-1. Atmosphere is divided into 5 major layer or sphere.

1. Troposphere This is the layer closest to the earth. It has 75% of atmospheric gases, dust, water vapors and clouds. All major weather changes on earth are caused by the troposphere. Troposphere extends 10 – 16 km. above the surface of earth. Temperature in this layer varies from 30^0C to -50^0C.

2. Stratosphere This layer lies above Troposphere and extends from 16 – 50 km. It is clear layer, free from dust, clouds etc. The temperature remains steady in this zone. Jets often fly in stratosphere. Ozone layer lying in this zone insulates earth against the harmful ultraviolet rays of the sun. Temperature in this layer increases with altitude i.e. from -50^0C to -20^0C.

3. Mesosphere This layer extends unto 80 km. from 50 km. This layer is extremely cold and temperature is -90^0C at height of 80 km.

4. Ionosphere In this layer temperature increases from -90^0C at 80 km to 1200^0C at 480 km. It has electrically charged gas particle known as ionosphere. Ionosphere reflects radio waves.

5. Thermosphere This layer extends from 80 km to 480 km. Ionosphere lies in this layer.

6. Exosphere Exosphere begins at 480 km. and extends unto 700 km. and fades in vacuum / space. Hydrogen is the main gas present in this layer.

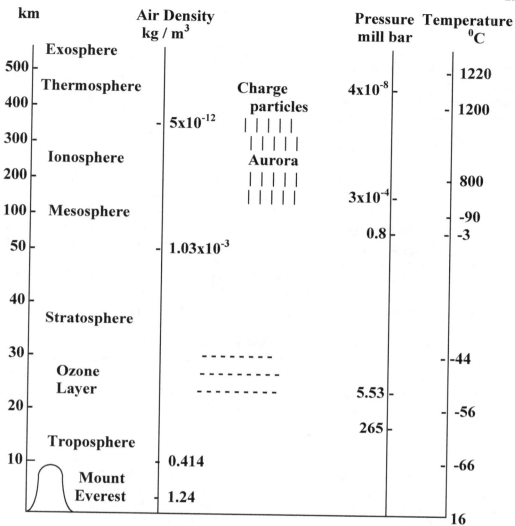

Figure-1: Structure of atmosphere

2.3 PARTICLES IN ATMOSPHERE

The atmosphere is never free from particles i.e. haze or haze aerosol. The scattering from gas molecules gives deep blue color to sky, while haze gives gray hue color to sky. The suspended particles size in atmosphere varies from 0.01 to 10 μm. Ordinary dust particles are non-hygroscopic, but salt bits from sea are highly hygroscopic and are responsible for condensation of water vapors into drops whose size depends upon atmospheric temperature and humidity. Haze has more concentration near earth's surface. But numbers of particles are found in stratosphere due to ash of metros, volcanic ash, and combustion from supersonic jet aircraft and air convention currents. However processes known as coagulation constantly remove particles from atmosphere, followed by washout. The large number of particles have radius of about 0.07 μm in haze [5].

2.4 FOG, CLOUDS AND RAIN

Fog droplets develop when the air layer near the ground becomes saturated. Haze particles extend up to an altitude of several thousand meters, while fog layer is only few hundred feet thick over the ground. The term advection refers to the horizontal motion of air. Advection fog is formed when moist and warm air moves across water or land having lower temperature. Such fog is common on the ground banks. Air with a long track over hot Gulf stream blows across the cold Babrador current, making this as one of the world largest foggiest region. Radiation fog is formed when the ground losses heat at night by radiation through a clear atmosphere and chilling the overlapping moist air. Fog of this class occurs frequently in inland area and may be intensified by cold air drainage from sloping terrain. Advection-radiation fog is formed by a combination of the two processes and is common ground lake region and the Atlantic coastal plane. Evaporated fog, often called warm water fog is produced, when the vapor from a water surface rises into colder, comparatively quiet air, depending on the temperature difference, a stable layer of day fog then develops. Fog of this type often observed over lakes and stream during autumn morning, especially when an overnight cold snow has occurrence. Frontal fog is formed when warm air falling through a layer of cold air i.e. made of warm moist air slides over and mingles with a colder surface layer i.e. temperature inversion is required.

Size of fog droplets grows continuously and sufficient amount of lantern heat of vaporization is released due to condensation to reverse the temperature trend. The repeated condensation may correspond to the transit condition called mist. The size grows to 8 μm and when fog is fully evolved, most of its droplets are of 12 μm, though sizes of droplets vary from 3 μm to 70 μm. Droplets concentration is 1 to 10 drops cm^{-3} and water content of 0.25 to 0.5 gm m^{-3}.

Cloud droplets are similar to fog, but at higher altitude i.e. when ground air rises and becomes saturated with water vapor as it cools at high altitude and becomes fog. Due to presence of hygroscopic particles in the atmosphere, clouds are formed, even when saturation point is not reached. The size of droplets grows due to collision process. The size of droplets is 0.75 to 2.25 mm, rain fall rate is 1mm h^{-1}, 1.25 to 3 mm for rain 5mm h^{-1} to 4.5 mm when rate of rain is 25 mm h^{-1}. Water contents or number of drops are about 1000 m^{-3} [5].

2.5 SCATTERING OF LASER BEAM

Scattering is the process by which particles in the path of an electromagnetic wave continuously abstract energy from the incident wave and reradiate that energy into the total solid angle at the particle, some of the energy is converted into heat or is reradiate at different frequency. If the particle size is smaller than one-tenth of the wavelength of light, then the scattering is known as Rayleigh scattering. In case particle size is greater than one-tenth of wavelength of light then relation given by Mie gives the scattering and is known as Mie scattering. If the frequencies of scatter light changes, then the scattering is known as Raman scattering.

The particles responsible for atmospheric scattering are air molecules-radius 10^{-4} μm, concentration 10^{19} cm^{-3}; dust particles-radius 10^{-3} to 10^{-2} μm, concentration 10^4 to 10^2 cm^{-3}; haze particles-radius 10^{-2} to 1 μm, concentration 10^3 to 10 cm^{-3}; fog droplets-radius 1-10 μm, concentration 100 to 10 cm^{-3}; cloud droplets-radius 1-10 μm, concentration 300-10 cm^{-3}; rain drops-radius 10^2 to 10^4 μm, concentration 10^{-2} to 10^{-5} cm^{-3} [7,8].

2.5.1 Rayleigh Scattering

The basis of Rayleigh scattering is that when electromagnetic wave interacts with gas molecules or molecules of a particle, the negative and positive charges are separated and induced dipole is created from non-polar molecules. This dipole oscillates with electromagnetic wave, abstract energy and sends it up wave in space that proceeds outward as though the primary wave were not present. The scatter energy from incident laser beam due to gas molecules is available equally in forward and backward hemisphere as shown in figure-2a. The scatter light is depolarized and the expression for gas molecular scattering coefficient is given by

$$\sigma_R(\lambda) = \frac{8\pi^3(n^2-1)^2}{3N\lambda^4} \cdot \frac{6-3\delta}{6-7\delta} \qquad \text{...............[1]}$$

Where N is the number of molecules per unit volume, n is the refractive of the medium, λ is the wavelength of incident radiation, δ is the depolarization factor of scatter radiation and $\delta = 0.035$ [6].

The Rayleigh scattering coefficient depends inversely on volume concentration of gas molecules and inversely to fourth power of wavelength. For example transmission loss due to molecular scattering of vertical column of entire atmosphere is 9.1% and 0.7% for wavelength 0.55 and 1.06 mm respectively [6].

The total scattering cross section σ_p of a small particle having radius r, which is less or equal to 0.03, is given by

$$\sigma_p = \frac{128\,\pi^5 r^6\,(n^2-1)^2}{3\,\lambda^4\,(n^2+2)^2} \qquad \text{......[2]}$$

The scattering efficiency factor Q_{sc} defined as the ratio of the total scattering cross section to the geometric cross section of the particle is given by

$$Q_{sc} = \frac{128\,\pi^4 r^4\,(n^2-1)^2}{3\,\lambda^4\,(n^2+1)^2} \qquad \text{........[3]}$$

Thus the scattering efficiency factor of small particle is proportional to fourth power of ratio of radius of particle to wavelength of light and is dimensionless quantity.

2.5.2 Mie Scattering

When the particle size is greater than 0.03 times the wavelength of light then the scattering does not follow Rayleigh scattering pattern since scattering is more in forward direction with maxims and minims appears at various angles as shown in figure-2b & 2c, the theory of large scattering pattern i.e. scattering pattern of colloidal particle, considering them as isotropic spheres was first developed by Mie in 1908. Later, Stratton [8] presented theory on mathematical basis.

The basis of Mie theory is on oscillating multi-poles i.e. when the particle size increases, particle consist of many closely packed complex molecules considered as array of multi-poles. Under the influence of incident beam field, these multi-poles oscillate in synchronization with incident beam. The oscillating multi-poles give rise to secondary electromagnetic waves called partial waves, which combine in the far field to give rise to scatter wave. The partial waves are represented in the theory by successive amplitude terms in a slowly converging series, whose squared submission gives the scattered intensity at a particular observation angle. Because the size of particle is comparable with wavelength, phase of primary wave is not uniform over the particle, resulting in spatial and temporal phase difference between the partial waves. The phase difference in partial waves causes interference causing a complex pattern in scatter wave with maximum and minimum in intensity depend upon the distance from scatter particle and its direction. The total amount of light scattered in all direction is given by scattering cross section of particle and is related to its geometrical cross section. The attenuation of beam depends on concentration and cross section of the particles. For detail theory, reader can read in book as mentioned earlier.

The intensity of scatter beam depends on relative size parameter, α ($2\pi r/\lambda$), of particle, and direction θ and complex refractive index m of particle.

Where $m(\lambda) = n(\lambda) - i \, n_i(\lambda)$ ....[4]

$n_i(\lambda)$ depends on absorption of light.

The intensity $I(\theta)$ of scatter beam in direction θ of incident wave is given by,

$$I(\theta) = E \, \frac{\lambda^2}{4\pi^2} \cdot \frac{(i_1 + i_2)}{2} \qquad \qquad ....[5]$$

Where E is intensity of incident beam on particle of radius r, and λ its wavelength.

$$i_1(\alpha, m, \theta) = |S_1|^2 = \left| \sum_{n=1}^{\infty} \frac{2n + 1}{n(n+1)} (a_n \, \pi_n + b_n \, \tau_n) \right|^2 \qquad ......[6]$$

$$= |Re(S_1) + Im(S_1)|^2 \qquad \qquad ....[7]$$

$$i_2(\alpha, m, \theta) = |S_2|^2 = \left| \sum_{n=1}^{\infty} \frac{2n + 1}{n(n+1)} (a_n \tau_n + b_n \pi_n) \right|^2 \qquad \qquad ....[8]$$

$$= |Re(S_2) + Im(S_2)|^2 \qquad \qquad ......[9]$$

S_1 and S_2 are dimensionless complex amplitudes observed as intensities when the particle is illuminated by plane-polarized light whose electric vector is perpendicular and parallel to the plane of observation. The vertical line indicates that the absolute values of the complex argument are to be taken. a_n and b_n are amplitude of electric and magnetic wave of nth dipole of particle. Each intensity function is found as the sum of an infinite series. Each series converge slowly, and where α is greater than unity, the number of term requirement for satisfactory convergence is some what greater than the value of α.

Incident Beam   **Small Particle**

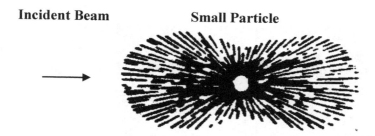

Figure-2a: Particle size 1/10 wavelength or less

Large Particle

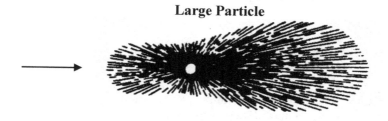

Figure-2b: Particle Size 1/5 Wavelength of Light

Large Particle

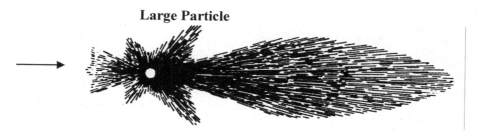

Figure-2c: Particle Size Larger than Wavelength of Light

The values of a_n and b_n are found from Ricatti-Bessel functions, whose arguments are found from the particle relative dimension α complex refractive index m. Extensive tables of Ricatti-Bessel functions for scattering applications have been published for wide range of particle size and refractive index for spherical particles [9 to 13, 34]. The functions π_n and τ_n depends only on the angle θ and involves the first and order derivative of Legendre polynomials having order n and argument $\cos\theta$, these functions are tabulated and reviewed [9]. The symbol Re and Im denotes the real and imaginary part of S_1 and S_2. Thus from these data $i_1(\alpha,m,\theta)$ and $i_2(\alpha,m,\theta)$ can be found. Detail instructions for computing the functions are given in books and publications [14,15].

2.5.3 Total Scattering

The total scattering σ_p by particle in all direction is the amount of flux scattered from incident flux and generally expressed as efficiency factor Q_{sc} related to geometrical cross section of particle i.e.

$$Q_{sc} = \sigma_p / \pi r^2 \qquad \ldots\ldots\ldots[10]$$

Total scattering cross section after integrating the scatter flux by particle in all direction is given by

$$\sigma_p = \frac{\lambda^2}{2\pi} \sum_{n=1}^{\infty} (2n+1)(|a_n|^2 + |b_n|^2) \qquad \ldots\ldots\ldots[11]$$

Scattering efficiency factor of particle is given by

$$Q_{sc} = \frac{2}{\alpha^2} \sum_{n=1}^{\infty} (2n+1)(|a_n|^2 + |b_n|^2) \qquad \ldots\ldots\ldots[12]$$

When α is small Q_{sc} is less than unity, rises to maximum value of 4 when wavelength of light is equal to radius of water particle, it converges to value of 2 like damped oscillation as radius increases, at absorption wavelength of light it converges to value 2 only like critical damped oscillations, shows diffraction by particle is equally responsible in attenuating light.

2.5.4 Volume Scattering Coefficient

The total amount of flux removed from beam is just N times that are removed by one particle i.e.

$$\beta_{sc} = N \sigma_p \qquad \ldots\ldots\ldots[13]$$

Where N is particle concentration.

2.6 ABSORPTION SPECTROSCOPY OF ATMOSPHERIC GASES

A molecule possesses energy that is defined by its various movements like its translation, vibration, rotation, and its electron orbit with spin and nuclear spin. Its total energy E_T can be expressed as:

$$E_T = E_{TR} + E_{EL} + E_{VIB} + E_{ROT} + E_{SPIN} \qquad \ldots\ldots[14]$$

Where E_T, E_{TR}, E_{EL}, E_{VIB}, E_{ROT} and E_{SPIN} are its energies associated with its transition, electronic, vibration, rotation and nuclear spin movement.

The energy states between the ground electron states and the first excited state of most of molecules is more than 4 electron volt (eV), which means that resonance wavelength due to its electron excitation lies in ultra violet region of spectrum. While its energy difference between first and second vibration states of molecules is in tenths of eV, therefore its resonance absorption for vibration state change lies in mid infrared region of light spectrum. Its rotational energy difference between its rotational level is tens of thousands eV, therefore its adsorption or emission due to its rotational energy level change is in far infrared or in microwave region. Its transnational energy can take any value, hence responsible for Doppler broadening of spectral lines, nuclear spin is very slow and it has no effect on molecular spectra.

A diatomic molecule comprising on two unlike atoms, possesses a dipole moment and therefore can emit or absorb electromagnetic energy when it undergoes a transition from one vibration state to another. The quantum selection rule for vibrational transitions in such a heteronuclear molecule takes the form.

$$\Delta v = \pm 1 \qquad \ldots\ldots\ldots\ldots[15]$$

Change in the vibrational quantum number is restricted to a value ± 1. The frequency emitted or absorbed in a transition of this kind is given by

$$h\nu = E_{v'} - E_{v''} \qquad \ldots\ldots\ldots\ldots[16]$$

Where v' and v'' are the vibrational quantum numbers of upper and lower vibrational levels.

In a homo-nuclear molecule like O_2 and N_2 there in no dipole moment at any value of inter nuclear separation, and consequently such molecules neither absorb nor emit infrared radiation, although they can be vibrationally excited or de-excited through collisions.

In case of a polyatomic molecule, there are many different modes of vibrations, this lead to complex band spectrum. Each nucleus of a free atom has three degrees of freedom, which is reflected in the total transition energy (3kT/2) of such an atom, since kT/2 represents the equilibrium mean thermal energy associated with each degree of freedom according to statistical mechanism.

Thus N – atom molecules has 3N degrees of freedom. Three of these degrees of freedom are required for the total transnational motion of the molecule, and three more (two in case of a linear molecule) are involved in the free rotation of the molecule. This leaves $3N - 6$ (or $3N - 5$ for a linear molecule) possible modes of vibration. Simple molecules like H_2O and CO_2 have three modes of vibration, designated v_1, v_2 and v_3, one of these can be thought of as a bending vibration, while the other two relates to symmetric and asymmetric vibration along the inter-nuclear axis.

The gases producing green house effect and harmful to living organism and their quantity is increasing in air due to industrialization and man's need are Carbon dioxide – CO_2, Carbon monoxide -CO, Nitrous oxide - N_2O, Nitrogen oxide - NO, Nitrogen dioxide -NO_2, Sulfur dioxide - SO_2, Chlorofluorocarbon - CFC-11 & CFC-12, Ozone – O_3, Methane – CH_4 and Water vapors – H_2O. The two gases Nitrogen and oxygen present in the highest concentration in the atmosphere are homo-nuclear do not exhibit molecular absorption lines/band.

2.6.1 Absorption Spectra of Ozone O_3

Ozone is formed due to dissociation of O_2 molecule by ultraviolet radiation from sun. Ozone has three fundamental vibration-rotation absorption bands in the infrared region 9.1, 14.1 and 9.67 and other bands with center at 5.75, 4.75, 3.95, 3.27 and 2.7 μm width of each band is of the order of 0.1μm. The strongest absorption band is at 4.75 μm. Atmosphere has maximum concentration of Ozone is at height of 22 kilometer from sea level, where maximum ultraviolet radiation is absorbed by Oxygen. [6]

2.6.2 Absorption Spectra of Water Vapors H_2O

The vibration-rotation spectra of water vapors fill the entire visible region as well as near and mid infrared region unto 10.0 μm. Visible and near infrared have weak absorption lines, while mid infrared has strong absorption bands. Near infrared absorption bands have center at 0.72, 0.81, 0.94, 1.1, 1.38 and 1.87 μm, while mid infrared are at 2.66, 2.74, 3.17 μm and 6.27 μm is strongest and broadest absorption band. In vertical column of air, solar radiation is completely absorbed from 2.6 to 3.3 μm and 5.5 to 7.5 μm [6].

2.6.3 Absorption Spectra of Carbon dioxide CO_2

Carbon dioxide molecule has four fundamental vibration frequencies. The fundamental vibration-rotation v_2 band with center around 15 μm with 14 hot bands occupies a broad interval of spectrum from 12 to 20 μm. In the center region i.e. 13.5 to 16.5 μm vertical column of the atmosphere completely absorb solar radiation in this band. Next fundamental vibration-rotation absorption v_3 band induces very strong absorption in the atmosphere, known as 4.3μm absorption band. The intensity of the 4.3

μm band is so great that solar radiation is completely absorbed in the wavelength from 4.2 to 4.4 μm due to vertical column of atmosphere. Beside 15 and 4.3 μm band 10.4, 9.4, 5.2, 4.7, 2.7, 2.0, 1.6 and 1.4 μm of light is also absorbed by carbon dioxide with a fainter band in the region 1.24 to 0.78 μm. All these bands have bandwidth of 0.1μm [6]. Absorption in mid, near and far infrared is mainly due to molecular absorption band of CO_2 and H_2O. Figure-5 shows atmospheric transmittance over a path at sea level for length 1880 meters showing absorption in atmosphere is due to resonant molecular absorption by H_2O and CO_2 present in air.

2.6.4 Absorption Spectra of Methane CH₄

The methane molecule represents a tetrahedron with spherical top type. Due to symmetry of molecule, it has only two-absorption lines 3.3 and 7.7 μm and maximum number of overtones and combination frequencies in the interval of 1.6 to 3.9 μm. This gas has maximum absorption at 3.391 and 3.270 μm [6].

2.6.5 Absorption Spectra of Carbon Monoxide CO

The fundamental vibration-rotation band of CO molecule is situated near 4.67 μm. It has other infrared absorption bands 1.19, 1.57, 2.35, 2.36, 2.41, 2.47, 2.5 and 4.67 μm. Electron transition occurs at frequency less than 1 μm. The pure rotational spectrum of the CO molecule is situated in the far infrared regions of the electro-magnetic wave scale.

2.6.6 Absorption Spectra of Sulfur Dioxide SO₂

The gaseous sulfur dioxide molecule belongs to the class of asymmetrical tops. Its molecules have relatively strong absorption band with center around 4.001, 8.880, 9.024 and 19.3 μm [6].

2.6.7 Absorption Spectra of Nitrous Oxide N₂O

The nitrous oxide molecule is linear asymmetrical substance with strong electron band in the far ultraviolet. Its fundamental vibration frequencies 7.8, 17.0 and 4.6 μm are octave in the infrared spectra.

2.6.8 Absorption Spectra of Nitrogen Dioxide NO₂

Nitrogen dioxide molecule is of the asymmetrical top type. It has fundamental absorption band in infrared. Center positions of its three fundamental absorption bands are 7.58, 13.34 and 6.18μm. Their center positions of 22 overtones and combination frequency band in the wavelength interval are from 1.6 to 6.7 μm.

Table1: Concentration of Atmospheric Gases with Fundamental Vibratio-Rotational Absorption Lines [3, 5, 6, 18]

Gases	Volume Concentration	Fundamental Vibration Rotational Absorption Line
1. Nitrogen	78.084%	No Vibration-Rotation Absorption Lines
2. Oxygen	20.946%	No Vibration-Rotation Absorption Lines
3. Argon	0.934%	No Vibration-Rotation Absorption Lines
4. Carbon dioxide	0.0314%	4.3 and 15.0 μm
5. Water Vapor	0 - 2%	2.66, 2.74 and 6.25 μm
6. Nitrogen dioxide	2 ppm	6.25, 7.58 and 13.34 μm
7.Sulfur dioxide	0 – 20 ppb (More in urban area)	7.3, 8.7 and 19.5 μm
8.Ozone	0.07 ppm (ground) 1-3 ppm (22km height)	9.0, 14.1 and 9.6 μm
9. Methane	2 ppm	3.3 and 7.7 μm
10. Carbon monoxide	1.1 ppm	4.67 μm.
11. Nitrous oxide	0.5 ppm	4.6, 7.8, and 17.0 μm
12. Chlorofluorocarbon CFC-11 CFC-12 CFC-113	250 ppt 550 ppt 80 ppt	9.220 and 11.806 μm 10.860, 10.834 μm 9.604 μm

2.6.9 Absorption Spectra of Chlorofluorocarbon (CFC)

Main absorption lines of this man made compounds now presents in air are: CFC-11: 11.806, 9.220, CFC-12: 10.860, 10.834 and for CFC-113: 9.604 μm

2.7 TRANSMISSION OF LASER BEAM

As the laser beam at particular wavelength λ travels through the atmosphere, its intensity reduces exponentially with distance R given by relation

$$I_R(\lambda) = I_0(\lambda) \; e^{-\alpha(\lambda)R} \qquad\qquad \text{................[17]}$$

Where $I_R(\lambda)$ and $I_0(\lambda)$ are the radiation intensity before and after transmission of laser beam through distance R in the atmosphere, $\alpha(\lambda)$ R is also called optical thickness of the medium, $\alpha(\lambda)R$ depends on scattering and absorption from molecules of air, trace gases, water vapor and aerosol, if the wavelength of laser beam coincide with the absorption spectral lines of atmospheric gases or water vapors. It also depends on scattering from molecules of air gases and water vapor, scattering and absorption from suspended particles in air and scattering and absorption from haze/fog/rain. Attenuation is also represented in terms of transmittance τ defined as

$$\tau = \frac{\text{Emergent power}}{\text{Incident power}} = \exp(-\alpha R) \qquad \ldots\ldots[18]$$

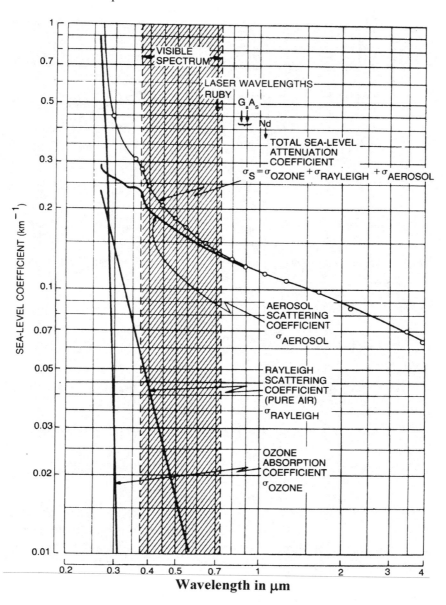

Figure-3 Calculated Atmospheric Attenuation Coefficient under Clear Atmosphere [36,37].

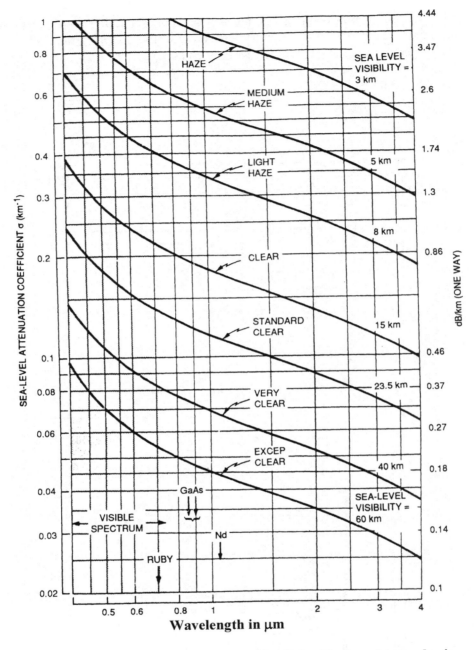

Figure-4 Variation of attenuation with wavelength for Various Atmospheric Conditions, neglecting resonant absorption due to water vapor and carbon dioxide [36, 37].

For 1 km path relation between attenuation and transmittance

$$\tau \ \ km^{-1} = antilog \ (- \ 0.4343\alpha) \ \ km^{-1} \qquad\qquad\ldots\ldots..[19]$$

$$\alpha \ km^{-1} = -2.303 \log \tau \quad km^{-1} \qquad\qquad ....[20]$$

Figure-3 gives calculated atmospheric attenuation at sea level under clear atmospheric condition from 0.2 to 4 μm, while figure-4 gives attenuation of light beam under various atmospheric conditions neglecting attenuation due to water vapor, carbon dioxide and other trace gases. Figure-5 gives actual transmittance of light beam with wavelength from 0 to 14 μm.

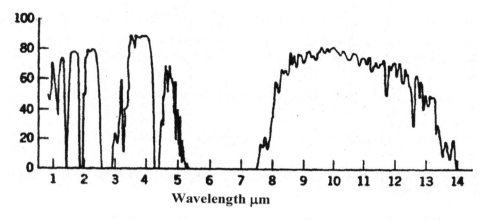

Figure-5: Percentage Transmittance Through the Earth Atmosphere Horizontal Path at Sea Level for path length 1828 meters [23]

2.7.1 Transmission of light through Haze, Fog, Clouds and Rain

As the particle size increases, the scattering coefficient, hence transmission of light does not depend much on wavelength of light. Through haze, visible wavelength transmission is almost uniform with wavelength, while through fog and clouds. Light is equally attenuated even in near and mid infrared. It has been observed [35] that from transition condition between haze and fog known as mist, the relative attenuation as compared to green i.e. beyond $\beta_{sc} = 6$ km^{-1}, the relative attenuation for blue (0.483μm) and red (0.675μm) light remains constant within 2% of the value for green (0.565μm) as shown in figure-6. Figure-7 gives light attenuation in visible and infrared for four type of fog. In rain, transmission in ultraviolet, visible, near infrared and mid infrared does not depend on wavelength, rather it has been observed that the transmission of light, relative attenuation of light increases more rapidly at 10 μm as compared to 1μm with increase in intensity of rain [5,6].

2.8 VISIBILITY

The term visibility is commonly used in stating how well we can generally see under given meteorological conditions.

Visual and Meteorological Ranges: Visual range [25] is the distance of specified target at that point background is just equal to the threshold contrast of an observer during daytime. It depends on atmospheric extinction. Target is generally a building against sky. The visual range R_v is given by formula

$$R_v = \frac{1}{\alpha_{sc}} \ln \frac{C}{\varepsilon} \qquad \qquad \dots\dots\dots\dots[21]$$

Where C is the inherent contrast of the target against the background, and ε is the threshold contrast of the observer. The use of α_{sc} rather than α_{ex} in implies that absorption by atmospheric particles at visual wavelengths is small to ignore. Meteorological range R_m target is specified as black target and $\varepsilon = 0.02$ [27], C=1 against sky background, R_m becomes

$$R_m = \frac{1}{\alpha_{sc}} \ln \frac{1}{0.02} = \frac{3.912}{\alpha_{sc}} \qquad \qquad \dots\dots\dots\dots[22]$$

Table-2 gives value of the meteorological range and scattering coefficient for the indicated meteorological conditions.

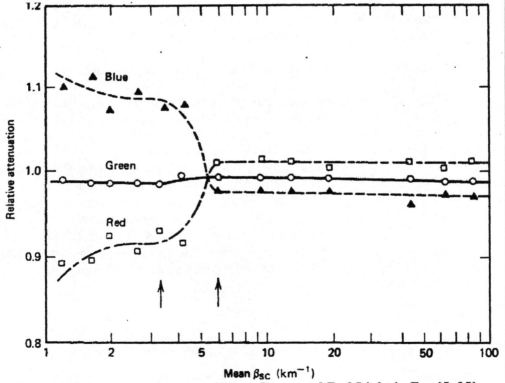

Figure-6 Relative Attenuation of Blue, Green and Red Light in Fog [5, 35].

43

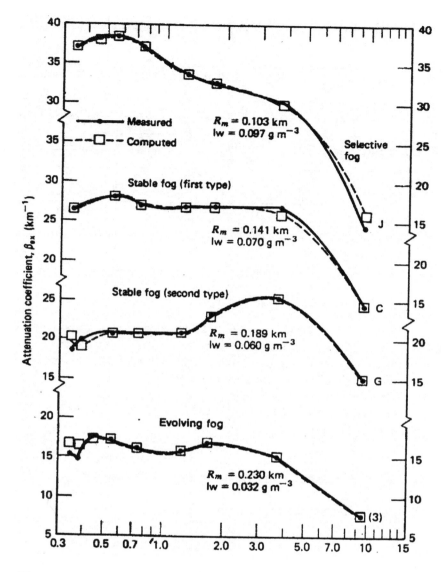

Figure-7- Spectral Attenuation of four type of fog [5,18,35]

2.9 RANDOM REFRECTION

The collimation of laser beam suffers due to refraction from water droplets when beam travels through fog, rain, and clouds or due to air turbulence and temperature inversion over a long distance. Time of travel of beam increase due to its path deviation as a result of temperature inversion, air turbulence, refraction or slant path terrestrial refraction [3]. This also results in broadening of laser echo after laser pulse travels to and fro from range finder to target and back.

Table-2: Scattering Coefficient and Meteorological Range under different Weather Conditions. [24]

Weather Conditions	Meteorological Range R_m in Meters	Scattering Coefficient α_{sc} (km^{-1})
Dense Fog	< 50 m	> 78.2
Thick Fog	200 m	19.6
Moderate Fog	500 m	7.82
Light Fog	1000 m	3.91
Thin Fog	2 km	1.96
Haze	4 km	0.954
Light Haze	10 km	0.391
Clear	20 km	0.196
Very Clear	50 km	0.078
Pure Air	277 km	0.0141

2.10 DISPERSION

Due to change in refractive index with wavelength, variations of maximum 6% in refractive angle are observed over the wavelength range from 0.4 to 11 μm [3]. Due to temperature inversion over long path, there is deviation of laser beam, when laser beam and aiming sight are in different wavelength range.

2.11 PROPAGATION of LIGHT BEAM under TEMPERATURE INVERSION CONDITION [16].

Consider a parallel beam propagating in the x-direction in a medium with refractive index n(z). Assume n(z) changes linearly with the height z.

During the interval dx, the beam is deflected by an angle

$$d\theta = \frac{dn}{dz} \cdot dx \qquad \qquad \dots\dots[23]$$

Where it has been assumed n ~ 1 and changes only slightly across the beam

The deflection of the beam at range R is then

$$d = \int_0^R x.d\theta = -\frac{1}{2}(dn/dz)R^2 \qquad \qquad \dots\dots[24]$$

provided (dn / dz) remains constant despite beam deflection.

The refractive index at pressure P and temperature t, from the Lorentz-Lorentz relation, can be written in the form

$$\frac{n^2 - 1}{n^2 + 1} = \frac{2}{3} A(\lambda) \, \rho(P,T) \qquad\qquad \text{------------[25]}$$

Where ρ is the density of air and λ is the wavelength, if $n \sim 1$, this reduces to

$$n - 1 = A(\lambda) \, \rho(P,T) \qquad\qquad \text{...............[26]}$$

Hence $\quad \dfrac{\delta n}{\delta T} = - A(\lambda) \dfrac{\rho}{T} \qquad\qquad \text{...........[27]}$

At constant pressure, assuming the ideal gas law from equation [24] and [27], we then obtain

$$d = \tfrac{1}{2} R^2 A(\lambda) \frac{\rho}{T} \cdot \frac{\delta T}{\delta z} \qquad\qquad \text{.............[28]}$$

In the optical and near IR region of the spectrum, the function $A(\lambda)$ is of the form

$$A(\lambda) = A_\infty \left[1 + \frac{B}{\lambda^2} \right] \qquad\qquad \text{...................[29]}$$

From the "Handbook of Geophysics and space environment" Mc Graw Hill 1965 page 9-1, we deduce

$$A(\lambda) = 0.2199 \left[1 + \frac{7.52 \times 10^{-3}}{\lambda^2} \right] \text{cm}^3/\text{gm} \qquad\qquad \text{......[30]}$$

Where λ is in μm

Hence equation [28] and [30] give

$$d = \tfrac{1}{2} R^2 \frac{\rho}{T} \, 0.2199 \left[1 + \frac{7.52 \times 10^{-3}}{\lambda^2} \right] \frac{\delta T}{\delta z} \qquad\qquad \text{.........[31]}$$

or in terms of pressure P (in m bars)

$$d = \tfrac{1}{2} R^2 \frac{P}{T^2} \, 7.76 \times 10^{-5} \left(1 + \frac{7.52}{\lambda^2} \right) \frac{\delta T}{\delta z} \qquad\qquad \text{.....[32]}$$

Taking P = 1000 m bar $\rho = 1.18 \times 10^{-3}$ gm / cm^3 corresponding to T = 300^0 K

$$d = 4.32 \times 10^{-7} R^2 \left[1 + \frac{7.52 \times 10^{-5}}{\lambda^2} \right] \frac{\delta T}{\delta z} \qquad \qquad \ldots [33]$$

At, $\lambda = 0.5 \ \mu m$, we then obtain from equation [33]
$$d = 4.45 \times 10^{-7} R^2 (\delta T / \delta z)$$

At, $\lambda = 2.0 \ \mu m$, we obtain from equation [33]
$$d = 4.33 \times 10^{-7} R^2 (\delta T / \delta z)$$

The difference in the deflection between the two wavelengths is

$$\Delta d = 1.2 \times 10^{-8} R^2 (\delta T / \delta z)$$

Where R is in cm. and $\delta T / \delta z$ is in ^{0}K cm^{-1}

For example, taking R = 4 km, $\delta T / \delta z = 0.1 ^0$K cm^{-1} we obtain
$$\delta d = 192 \text{ cm.}$$

Thus, laser beam at wavelength 2.0 μm will be deflected by 192 cm. under above condition due to atmospheric dispersion if aiming sight is visible i.e. at 0.5 μm

SUMMARY

1. Loss of energy of laser beam during propagation in the atmosphere is more in ultraviolet region due to molecular scattering and less in infrared region.

2. Traces of gases like water vapor, carbon dioxide, ozone present in atmosphere attenuates laser beam more over certain bands in infrared region.

3. Haze aerosols always present in the atmosphere attenuates visible light much more as compared to infrared region under poor visible conditions.

4. When aerosol particle size increases, scattering cross section is less dependent on wavelength. Through fog, clouds and rain, light in ultraviolet, visible and infrared is almost equally attenuated.

5. Collimation of laser beam is affected due to air turbulence, precipitation or temperature inversion. Highly collimated beam may miss the target during temperature inversion condition, if laser and aiming sight are at different wavelengths

LIST of SYMBOL USED

$I_o(\nu)$ = Iinitial radiation intensity

ν = Frequency of light

R = Range travel by light

R_v, R_m = Visual and Meteorological Range

$I_R(\nu)$ = Radiation intensity at range R

$\alpha(\nu)$ = Attenuation coefficient of light

τ = Transmittance

ppm = Parts per million

ppb = parts per billion

ppt = Parts per trillion

$\sigma_R(\lambda)$ = Rayleigh scattering cross section of gas molecules

δ = Depolarization factor of scattered radiation

λ = Wavelength of light

N = Concentration of gas molecules/particles per unit volume

n = refractive index of air/integer

σ_p = Scattering cross section of particle

r = Radius of particle

$m(\lambda)$ = Complex refractive index of particle

n_i = imaginary part of refractive index

α = Particle size relative to wavelength

θ = Angle between incident and scattered beam

E = Intensity of incident beam

$I(\theta)$ = Intensity of scattered beam

$i_1(\alpha, m, \theta)$,

$i_2(\alpha, m, \theta)$ = Intensity of scattered beams with orthogonal direction of polarization

S_1 and S_2 = Dimensionless amplitudes of scattered waves with orthogonal direction of polarization

a_n and b_n = Amplitude functions of nth dipole of particle

π_n and τ_n = First and second order direction of Legendre polynomials having order n

Q_{sc} = Scattering efficiency factor

β_{sc} = Total volume scattering

P, T, ρ = Pressure temperature and density of air

C = Contrast of target

ε = Luminance-contrast threshold

α_{sc} = Attenuation coefficient due to scattering only

α_{ex} = Attenuation coefficient due to scattering and resonant absorption

APPENDIX-A

Resonant absorption lines of trace gases and their absorption coefficients in some typical resonant absorption lines

Species	Resonant lines/bands in μm	Resonant wavelength	Absorption cross section	Reference
CO_2	1.4, 1.6, 2.0, 2.76, **4.3,** 4.7 5.2, 9.4, 10.4, **15**	2.002445 2.066725	4.47* 0.938*	6. 19, 23
CH_4	1.6537, 1.7, 3.27, **3.31,** **7.6**	3.367482 3.313418	414* 1030*	6, 19, 23
N_2O	2.87, 3.9, **4.7, 7.78, 17.0**	3.8903	4.94*	6, 20, 23
O_3	2.7, 3.27, 3.95, 4.75, 5.75, **9.1, 9.6, 14.1**	0.2536	11.3**	6, 22, 23
CO	1.19, 1.57, **2.35,** 2.36, 2.41, 2.47, 2.5, **4.67,** 4.717, 4.81	4.709	2.8**	3, 6, 23
SO_2	3.3, 4.0, **7.32, 8.65,** 9.024, 9.6, **19.3**	0.300	1.3**	6, 17
NO_2	**6.18,** 6.2293, **6.25, 7.58,** **13.34**	0.4481	0.482**	6, 21
NO	**2.76, 5.3,** 5.263, 5.3, 6.069	5.215	0.67**	3
NH_3	2.087, 9.217, **6.1493,** 9.22 9.639, **10.333**	10.333	0.959**	18
CFC-11	9.24, **9.261,** 11.806	9.261	1.09**	3, 18
CFC-12	**10.719,** 10.834, 10.860	10.719	1.33**	18

* x 10^{-21} cm^2

** x 10^{-22} m^2

Wavelengths in bolts figure indicate fundamental modes of resonant absorption.

APPENDIX-B

Radiometric Quantity: Refer to narrow spectral band, as photon energy depends on light frequency.

Table-I: Dimension and Units of the Radiometric Quantities

Radiometric Quantities	Dimensions	Basic Unit
Radiant Energy	ML^2T^{-2}	Joules(J)
Radiant density	$ML^{-1}T^{-2}$	Joules per cubic meters (Jm^{-3})
Radiant Flux	ML^2T^{-3}	Watts (Joules per second $J\,sec^{-1}$)
Radiant Flux density at Surface		
Irradiance	MT^{-3}	Watts per square meter (Wm^{-2})
Radiant intensity	ML^2T^{-3}	Watts per steradian ($W\,sr^{-1}$)
Radiance	MT^{-3}	Watts per square meter per steradian

Photometric Quantities Refer to spectral band where human eye is most sensitive (555 nm). Sensitive curve of human eye is shown in figure 1B.

In photometry, the unit of Luminous flux is the lumens (lm). The lumen is defined as the luminous flux emitted into 1 sr by a source whose luminous intensity is 1 candela (cd).

The candela is defined as one – sixtieth of the luminous intensity of a blackbody radiator, having an area of 1 cm^2, at the temperature of solidifying platinum (2042K). at 555 nm

1 W = 680 lm

1 lm = 1.47×10^{-3} Watts

Table-II: Units and dimensions of Photometric Quantities

Quantity	Dimensions	Unit
Luminous Flux	ML^2T^{-3}	Lumen (lm)
Luminous Energy (quantity of light)	ML^2T^{-2}	Lumen second (lm sec) Talbot (T)
Luminous density	$ML^{-1}T^{-2}$	Lumens second per cubic meter (lm sec m^3)
Luminous Efficiency		Lumens per Watt
Luminous Flux Density at Surface		
Illumination	MT^{-3}	Foot Candle (lm ft^{-2})
Luminous Intensity (Candle Power)	$ML^2\,T^{-3}$	Candela (lm sr^{-1})
Luminous (Photometric Brightness)	MT^{-3}	Candela per unit area Lambert (cd $\pi\,cm^{-2}$) Foot-lambert (cd $\pi\,ft^{-2}$) Nit (cd m^{-2})

APPENDIX C

Molecular Quantities

A gram molecular weight, or mole, of any substance has a mass in grams equal to molecular weight. A mole of any substance contains the same number of molecules as a mole of any other substance. Thus a mole is a definite quantity of matter called a molar mass and denoted by M_A.

Avogadro's law states "Equal volumes of gas, under same conditions of temperature and pressure, contain equal numbers of molecules." The number of molecules in a mole of any gas is known as Avogadro's number N_A and has the value

$N_A = 6.025 \times 10^{23}$ Molecules

The volume of 1 mole of gas at 0^0C and 1 atmosphere of pressure is called molar volume V_A and has value

$V_A = 2.242 \times 10^4 \, cm^3 = 22.42$ litres

Loschmidt's number gives the number of molecules in 1 cm^3 of gas

$N_L = 2.687 \times 10^{19} \, cm^{-3}$

Response of the Eye

Photometry is the division of radiometry in which radiant flux is evaluated largely replaced the eye in photometric work, to obtain results that agree with visual result. Therefore eye is ultimate judge of light in the spectral region from about 0.38 to 0.76 μm. The human eye and associated neural system constitute the most remarkable optical sensor of all time. The capabilities of adaptation enable the eye to function at starlight levels of illumination as well at sunlight levels- dynamic range greater than 1 to 1 millions. Selective long time constant of eye (about 0.1 sec) produces a persistence of vision that makes possible the cinema and without which all motion viewed by lighting would exhibit intolerable stroboscopic effects. Finally, response to wavelength (colors) gives us additional information and immeasurably enriching our aesthetic appreciation. Peak response of eye shift to higher wavelength with increase in illumination [26] as shown in relative spectral response of eye as shown in figure-1B.

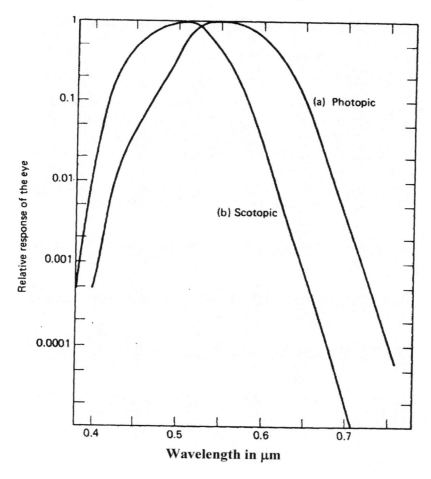

Figure-1B Spectral Response of Eye

REFERENCES

1. Herzberg,G., "Electronic Spectra and Electronic Structure of Polyatomic Molecules", Vol.III, VanNorstand, Princeton (1967).

2. Steinfeld,J.I., "Molecules and Radiation, An Introduction to Modern Molecule Spectroscopy", Publisher: Harper and Row, New York (1974).

3. Hinkely,E.D., Ku,R.T. and Kelly,P.L., "Techniques for Detection of Molecular Pollutants by Absorption of Laser Radiation", Laser Monitoring of the Atmosphere (E.D.Hinkely, Ed.), Topics in Applied Physics, Vol. 14, Publisher Springer-Verlag (1976).

4. Mansharamani,N., Greenhouse Effect-Lidar Techniques, (ISBN 978-81-7525-789-4) Publisher: Sita Publisher, Dehradun-248001, India (2006).

5. Earl J.McCartney, "Optics of the Atmosphere", John Wiley & Sons, New York & London (1976).

6. Zuev,V.E., "Laser Beam in the Atmosphere", Translated from Russian by Wood,J.W., Publisher- Consultants Bureau, New York and London (1982).

7. Victor Twesky, "Rayleigh Scattering", Appl. Optics (USA), Vol. 3, No. 10, October 1964, pp. 1150-1162.

8. Stratton,J.A., "Electromagnetic Theory", Publisher: McGraw Hill, New York (1941).

9. Van De Hulst,H.C., "Light Scattering by Small Particles", Wiley, New York, (1957).

10. Penndorf,R., "Review of Tables of Light Scattering", Part I, Appl. Optics (USA), Vol. 6, 1967, pp. 2019.

11. Penndorf,R., "Review of Tables of Light Scattering", Part II, Appl. Optics (USA), Vol. 7, 1968, pp. 1869.

12. Penndorf,R., "Review of Tables of Light Scattering", Part III, Appl. Optics (USA), Vol. 8, 1969, pp. 892.

13. Penndorf,R., "Review of Tables of Light Scattering", Part IV, Appl. Optics (USA), Vol. 10, 1971, pp. 2805.

14. Kerker,M., "The Scattering of Light and Other Electromagnetic Radiation", Academic Press, New York (1969).

15. Deirmendjian,D., "Electromagnetic Scattering on Spherical Polydispersions", American Elservier, New York (1969).

16. Fawcett,W. "Propagation of Light Under Temperature Inversion Condition", from Deputation Report of Mansharamani,N. (unpublished), Royal Signal & Radar Establishment, Great Malvern, U.K. (1977).

17. Brassington,D.J., "Sulfur dioxide Absorption Cross-Section Measurements from 290 to 317 nm", Appl. Optics (USA), Vol. 20, No.21, 1November 1981, pp. 3774-3779.

18. Mayer,A., et. al., "Absorption Coefficient of Various Pollutant Gases at CO_2 Wavelengths; Application at sensing of Pollutants", Appl. Optics (USA), Vol. 17, No. 3, 1 February, 1978, pp. 391-393.

19. Ambrico,P.F., et. al., "Sensitivity Analysis of Differential Absorption Lidar Measurements in the Mid-Infrared Region", Appl. Optics (USA), Vol. 39, No. 36, 20 December, 2000, pp. 6847- 6864.

20. Altman,J., Lahmann,W. and Weitkamp,C., "Remote Measurement of Atmospheric N_2O with a DF Laser Lidar", Appl. Optics (USA), Vol. 19, No. 20, 15 October, 1980, pp. 3453-3457.

21. Takeuchi,N., Shimizu,H. and Okuda,M., "Detectivity Estimate of the DAS Lidar for NO_2", Appl. Optics (USA), Vol. 17, No. 17, 1 September 1978, pp. 2734-2738.

22. Grggs,M., "Absorption Coefficient of Ozone in the Ultraviolet Region", J. Chem. Phys. Vol. 49, 1968, pp. 857-859.

23. Measures,R.M., "Laser Remote Chemical Analysis", A series of Monographs on Analytical Chemistry and its Applications, Vol. 94, John Wiley & Sons, New York, Toronto (1998).

24. Hulburt,E.O., "Optics of Atmospheric Haze", J. Opt. Soc. Am., Vol. 31, 1941, pp. 467-476.

25. Huschke,R.E., Glossary of Meteorology, AMS, Boston (1959).

26. Walsh,J.W., "Photometry", Dovers, New York (1964).

27. Blackwell,H.R., "Contrast Threshold of the Human Eye", J. Opt. Soc. Am, Vol. 36, 1946, pp. 624-643.

28. Arnulf,A. et. al., "Transmission by haze and fog in the spectral region 0.35 to 10 micron," J. Opt. Soc. Am., Vol. 47, 1957, pp. 491-498.

29. Green,A.E. and Griggs,M., "Infrared Transmission Through the Atmosphere", Appl. Optics (USA), Vol. 2, 1963, pp. 561-570.

30. Houghton,H.G., "The Transmission of Visible Light Through Fog", Phys. Rev., Vol. 38, 1931, pp. 152-158.

31. Kruse,P.W. et. al., Elements of Infrared Technology", Wiley, New York, 1963.

32. Kurnik,S.W. et.al., "Attenuation of Infrared Radiation by Fogs," J. Opt. Soc. Am., Vol. 50, 1960, pp. 578-583.

33. Meyer-Arendt,J.R., Radiometry and Photometry: Units and Conversion Factors," Appl. Optics (USA), Vol. 7, 1968, pp. 2081-2084.

34. Zelmanovicv,I.L. and Shifin,K.S., "Tables of Light Scattering". Publishing House, Leningard (1968).

35. Eldridge,R.G., "Mist-the Transition from Haze to Fog", Bull. Am. Meteorol. Soc., Vol. 50, 1969, pp. 422-426.

36. RCA, "Electro-Optics Handbook", RCA Corporation (1968).

37. Eppers,W., "Atmospheric Transmission", pp. 39-154, in Handbook of Lasers by Robert J. Pressley (Editor), Publisher CRC Press, Inc, Cleveland, Ohio, USA (1971).

CHAPTER-3

TYPE OF LASER SOURCES

3.1 INTRODUCTION

Ruby operating in red region i.e. 0.6943 µm is the first type of laser source [1] used for ranging [3] natural / non-cooperative targets, when R.W. Hellwerth [2] at Hughes Research Laboratories in 1961 produced giant optical pulse from ruby. At that time photo multiplier tube with S-20 response was used to detect very weak pulses in visible region after diffuse reflection from targets. Neodymium laser could not be used, since photo multiplier with S-1 response was noisy with very low quantum efficiency to detect signal at 1.06 µm. Even at that time lunar ranging was done with ruby source. Continuous wave (cw) helium-neon gas laser operating in red region with sinusoidal modulation was used to measure distance within millimeter accuracy by phase comparison with echo received from target fixed with retro-reflectors for survey purposes. Gallium Arsenide (GaAs) laser [17, 18] operating near infrared region could not be used for fieldwork as its operation was not possible at field temperature and silicon detector needed very low noise pre-amplifier for detection of weak laser echo signal. Therefore for quite sometime ruby and He-Ne lasers were used for ranging and distance measuring applications.

With the development of silicon avalanche photo-diode (Si APD) with internal carrier multiplication and good quantum efficiency in near infrared region of light and development of hetero structure Gallium Aluminum Arsenide (GaAlAs) semiconductor laser operating in pulse and cw mode [21] at room or field temperature with operating frequency at 0.82 µm matching to peak response of silicon photo-detector; neodymium glass and GaAlAs semiconductor lasers slowly replaced ruby and helium-neon (He-Ne) gas lasers for range and distance measurements. By mid seventies, almost all range finders, neodymium doped glass or crystalline materials were used. These sources had drawback i.e. range limitation in hazy atmospheric conditions and not safe for eye retina. With the development of pulse as well as frequency stable cw carbon dioxide laser, i.e. TEA CO_2 and cw wave-guide lasers operating at 10.6 µm found use in some ranging applications after development of cryogenic cooled mercury cadmium telluride (MCT) detector with peak response at this wavelength. Again with the development of Indium Gallium Arsenide InGaAs avalanche detector with maximum quantum efficiency at 1.5 µm, Raman shifted Neodymium doped Yttrium Aluminum Garnet (Nd:YAG) laser operating at 1.5 µm and Erbium (Er:Glass) doped glass lasers operating at 1.54 µm are used to range targets under eye safe conditions.

In last ten years, with dvelopment of sensitive detectors and compact thermal sights in mid infrared region operting at higher temperature, efficient nonlinear crystal and diode pumped solid state lasers, harmonic generation or optical parametric oscillators at 1.5 µm or in mid infrared are used in laser ranging system along with multiple role for target designation for terminal guidance of smart ammunition, depth measurement /

underwater detection or illumination for targets under adverse weather conditions. Further, if these sources are used for range finders, then these system can be modified to detect also pollutant in atmosphere [26].

In this chapter, solidstate lasers, TEA CO_2 lasers, Nd:YAG laser pumped optical parametric oscillators (OPO) and semiconductor lasers, junction and quantum cascades type which are the present state of art for ranging purpose along with designation of targets or illumination under adverse weather conditions are described.

3.2 REQUIREMENTS OF LASER SOURCE for RANGING

a. Its wavelength should have least attenuation under various atmospheric conditions.
b. Good target reflectivity at source wavelength.
c. Detector available at laser source wavelength should have fast response, good quantum efficiency, and low noise, rugged and operable under field condition without cooling.
d. Laser optics and optical elements for beam collimator and laser receiver should be cheap, having good transmission at laser wavelength.
e. Laser source energy can be stored in laser material for a single giant pulse.
f. Laser sources should be efficient, compact, cheap, long shelf and operating life and can operated in high repetition rate.
g. Operating electronics should be simple, cheap and less troublesome.

These requirements are compared in Table I with rating I-excellent, II- very good and III-good.

Table-I: Comparison of requirement of solid state, semiconductor and gas laser sources:

Laser sources: Requirements

Laser Sources	a	b	c	d	e	f	g	Remarks
Solid-state	II	I	I	I	I	I	I	Excellent
Semiconductors	II	I	I	I	III	I	I	Good
Gases	I	II	II	II	I	II	III	Very good
Nd:YAG Pumped OPO	I	I	I	II	I	I	I	Excellent

3.3 SOLID STATE LASER

3.3.1 Ruby Laser

Ruby [4, 5, 6, 8] consists of crystalline host Sapphire (Al_2O_3) in which a 0.05% of Al_2O_3 is replaced by Chromium oxide (Cr_2O_3) by weight percent. It is refractory material, hard and durable. It contains about 1.58×10^{19} Cr^{3+} ions / cm^3. Low concentration is preferred to avoid mutual interaction between active ions in order to keep fluorescence line width sharp. The pure single host crystal is uniaxial and possesses a rhombohedral or hexagonal unit cell. Very good quality crystal and in large size can be easily grown by Czochralski process with cylindrical axis in 0°, 60° or 90° configurations.

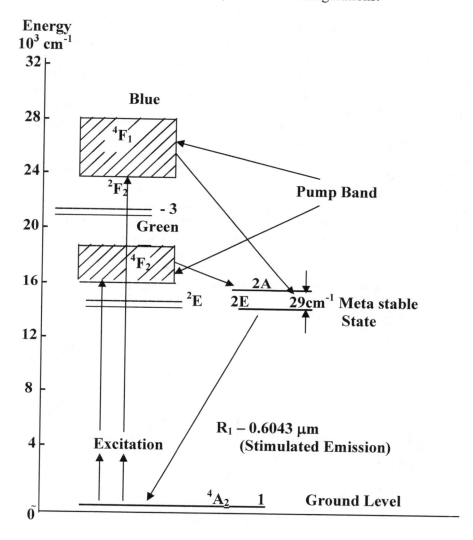

Figure-1: Energy Level Diagram of Cr^{3+} in Al$_2$O$_3$

Ruby is generally cut so that its cylindrical axis is at 60° to the optical axis. It has very good thermal conductivity at low temperature. Its ordinary and extraordinary indices of refraction are 1.764 and 1.756 at its lasing wavelength is 0.6943μm.

Spectroscopy of Cr^3 ion/its lasing properties:

Chromium is element from transition metal group. It has atomic number 24 with electronic configuration as

$$1s^2; 2s^2, 2p^6; 3s^2, 3p^6, 3d^5; 4s^1$$

Cr³⁺ has three d electrons in its unfilled shell, L=3, S=3/2. Its ground state is denoted by ^4F, Next state of Cr³⁺ion is ^2G with quantum numbers L=4, S=1/2. Ground level of free Cr³⁺ splits into three levels $^4F_1$, $^4F_2$ and $^4A_2$ as shown in figure-1. The next level, ^2G, splits into four sub levels designated by the symbols $^2A_1$, $^2F_1$, $^2F_2$ and ^2E. ^2E level splits into two levels that are only 29cm⁻¹ apart and are independent of crystal field and are sharp that is its width is 4.5 A°. The splitting of ^4F level depends strongly on crystal field with the result $^4F_1$ and $^4F_2$ levels are broad bands i.e. width of each level is 1000 A°. Thus the difference in width of these levels is of great importance for laser operation because broad level is needed for absorption and narrow one for emission. Lifetime in the excited state levels $^4F_1$and $^4F_2$ is extremely short while lifetime of meta stable level ^2E level is 3 millisecond that is ideal for giant pulse operation. The terminal level is ground level $^4A_2$. Main disadvantage with the ruby is that at least 50% population to be pumped from ground state $^4A_2$ to meta stable state ^2E for threshold. The quantum efficiency of ruby is 1%. Maximum upper state energy density is 4.52 J/cm³ and upper state energy at threshold is 2.18 J/cm³. The absorption spectrum of ruby laser crystal is shown in figure-2, its absorption spectrum matches well with emission spectra of xenon flash/mercury lamps. Very good quality beam with peak power can be generated from a single laser rod.

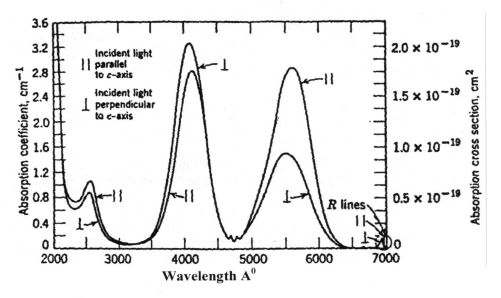

Figure-2: Absorption Spectrum of Cr³⁺ in Al₂O₃ (Ruby)

The ruby laser developed by author of this monograph at IRDE for range finder is shown in figure-3. It is air cooled with three plate sapphire resonant reflector with peak reflectivity of 86% is used as partial reflector. Total reflector is high quality fused silica prism with pyramidal error and right angle accuracy within 2 seconds of arc. Prism rotated at speed of 18,000 revolutions per minute and its position synchronized with xenon flash lamp using lamp and photodiode arrangement for triggering flash lamp.

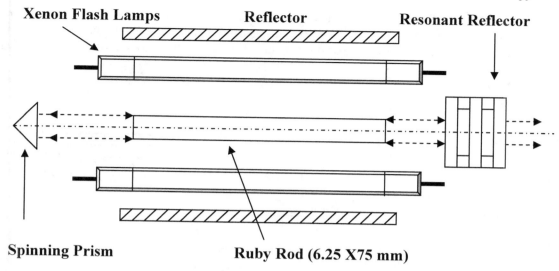

Xenon Flash Lamps **Reflector** **Resonant Reflector**

Spinning Prism **Ruby Rod (6.25 X75 mm)**

Figure-3: Q-Switched Ruby Laser

3.3.2 Nd:Glass Laser

In 1961, E. Snitzer [9] discovered laser action in glass doped with neodymium. (Nd) is one of the most important elements of rare earth element group to be used in laser range finders? Its lasing threshold is very low. Its operating wavelength, matches with most efficient, rugged and cheap silicon photo-detector. Natural targets reflectivity at its wavelength is good i.e. about 40%. At its operating wavelength 1.06 μm there is no molecular absorption due to atmospheric gases including water vapor, hence very good atmospheric transmission under normal visible condition. In glass Nd can be doped in very large quantity without affecting lasing property, giving quantum efficiency of 6% in xenon flash lamp pumping and 20% with semiconductor diode laser pumping. Its operating and shelf life is very large. Its life in meta stable state is large about 300 μs, therefore it can generate high peak power pulses with active and passive Q-switching methods. This laser material is most versatile for range finding. Nd can be doped in various types of glass host materials, but commercially it is available in silicate glass with concentration of 3% and in phosphate glass with concentration unto 7% as super-gain laser [12].

Spectroscopic and lasing property of Nd^{3+} ion in glass material:

Nd is from rare earth group element with atomic number 60 having electronic configuration as

$$1s^2; 2s^2, 2p^6; 3s^2, 3p^6, 3d^{10}; 4s^2, 4p^6, 4d^{10}, 4f^3; 5s^2, 5p^6, 5d^1; 6s^2$$

Its inner shell 4f is unfilled, while electrons are present in fifth and sixth sub-shell as $5s^2$, $5p^6$ & $6s^2$. Nd^{3+} ion has three electrons in the 4f-sub shell. The laser actions is due to excitation of electrons in 4f-sub-shell and are shielded due to presence of electrons in

60

the outer filled sub-shells which results in sharp meta stable state. At the same time excited state is broad i.e. results in broad absorption spectrum. In the ground state total orbital angular momentum (L) of three electrons in 4f-subshell are 6 atomic units and 3/2

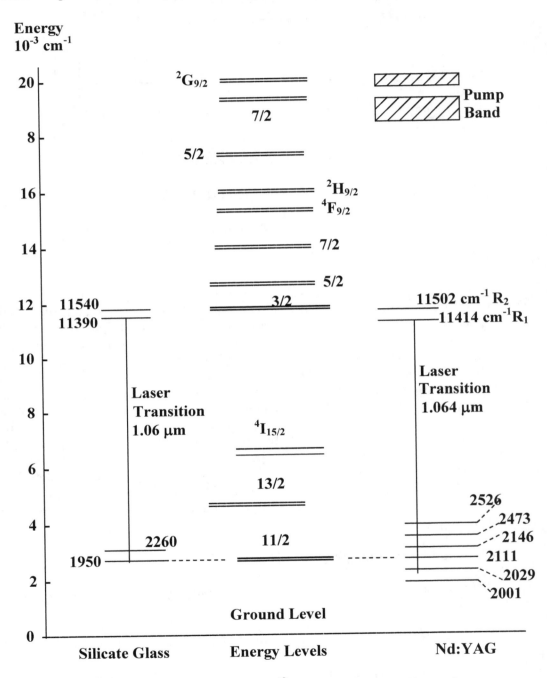

Figure-4: Energy Level Diagram of Nd^{+3} ions in YAG and silicate glass

or 1/2 atomic units due to spin angular momentum (S) depending on orientation of electron spin adds up vectorally, giving four values to total momentum of atom as $^4I_{9/2}$, $^4I_{11/2}$, $^4I_{13/2}$ and $^4I_{15/2}$. The first of these, which has the lower energy, is the ground level and other levels of Nd^{3+} ion and is shown in figure- 4. Laser action is generally from meta stable level $^4F_{3/2}$ to $^4I_{11/2}$ at 1.06 μm. Since life time of terminal level $^4I_{11/2}$ is very short and is 2000 cm^{-1} above ground level $^4I_{9/2}$; at room temperature this level remains empty. Therefore laser threshold of this material is very low of the order of 1 joule for xenon flash lamp excitation. At the same time lifetime of meta-stable level is about 300 μ sec. i.e. lot of energy can be stored in this material for giant pulse operation. The absorption spectrum of Nd:Glass is shown in figure -5. It has absorption at 0.58, 0.71 and 0.81 μm. Its lifetime in its absorption levels is extremely small and therefore excited atoms immediately come to meta-stable level.

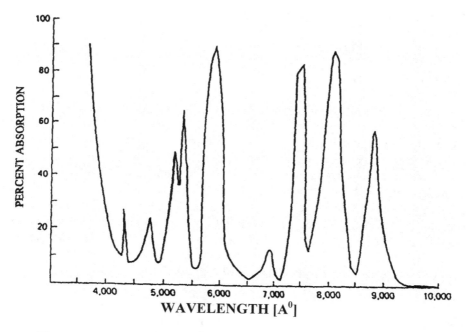

Figure-5: Absorption Spectrum of Nd Doped Silicate Glass

The Nd:Glass Laser developed at IRDE for laser range finder for armored vehicle is shown in figure-6. It is conventionally cooled system uses clad phosphate glass laser rod of 80 mm long with diameter 6.25 mm with core diameter of 4.5 mm, type Q-100 of M/S Kigre of USA [12]. The laser cavity consist of two plate sapphire resonant reflector with peak reflectivity of 66% as partial reflector and 90° porro prism rotated at a speed of 20,000 revolution per minute by DC permanent magnetic motor as total reflector. The laser rod is pumped by a Xenon flash lamp in a cylindrical Aluminum reflector. A small magnet embedded in prism mount induces an electrical pulse in a coil as its reference position for Q- Switching. Flash lamp is fired 95 μ second before the prism is aligned with resonant reflector. An 100 milli joules with 20-neno second duration pulse is obtained for ranging application.

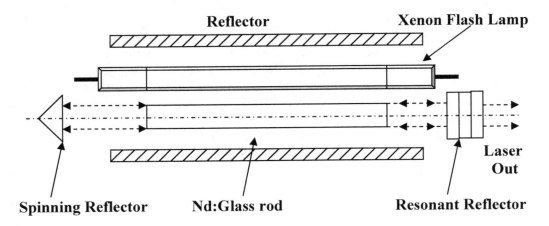

Figure-6 Nd:Glass Laser

3.3.3 Nd:YAG Laser

The Yttrium Aluminum Oxide $Y_3 Al_5 O_{12}$ (YAG) is known as Yttrium Aluminum Garnet [4,5,6,8] is crystalline material with cubic structure is very hard, have high melting temperature and having very good thermal conductivity with excellent optical quality. Since trivalent neodymium is doped in trivalent yttrium, therefore this material does not require charge compensation. Nd:YAG crystal has very slow growth rate and good quality of crystal are grown by Czochralski method. Nd concentration is generally more in centre; therefore large diameter boules are grown to fabricate laser rods of uniform doping from surrounding material. Due to inter action between Nd atoms energy levels, Nd atom percentage is limited between 1 to 1.5% i.e. 1.38×10^{20} atoms per cm^3.

Spectroscopic and Lasing Properties of Nd:YAG

Energy levels of Nd^{+3} ions in YAG crystal are shown in figure-4. It has narrow fluorescent line width of 4.5 A^o, which results in high gain of this material. Most preferred laser transition is for 1.064 μm, which originates from the R_2 components of the $^4F_{3/2}$ levels and terminates at the Y_3 component of the $^4I_{11/2}$ level. Its fluorescent lifetime is 230 μs, it can generate narrow pulses with high peak power. It has main absorption bands at 0.75 and 0.81 μm as shown in absorption   spectrum of Nd:YAG (figure-7). Its absorption band at 0.81 μm matches well with emission of GaAlAs semiconductor laser, therefore this material with laser diode pumping has highest overall efficiency as compared to other solid-state laser materials. Due to its low threshold, high gain and good thermal conductivity, this material is used in compact hand-held and high repetition rate range finders.

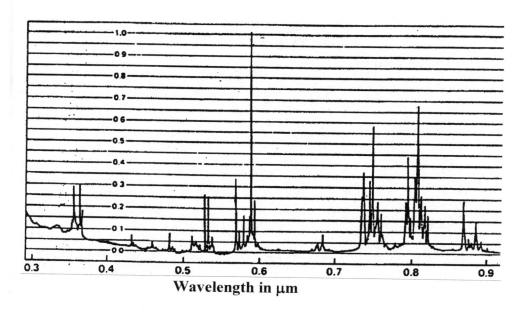

Wavelength in μm

Figure-7 Absorption Spectrum of Nd:YAG

Nd:YAG Laser developed at IRDE for hand-held range finder is shown in figure-8. It consist of Nd:YAG laser rod of 50 mm long and 5mm diameter with its one end coated for 50% reflectivity, while other end having anti-reflecting dielectric coating for 1.064μm. A passive Q-Switch acetate sheet with optical density of 0.4 at 1.06 μm of M/S Eastman Organic Chemicals of USA has been placed between total reflector and transmitting end of laser rod. The Nd:YAG rod is excited by Xenon flash lamp in a closely coupled silver coated cylindrical glass reflector pulse of 8 nanosecond duration with 20 mJ of energy is obtained for 8J input to flash lamp.

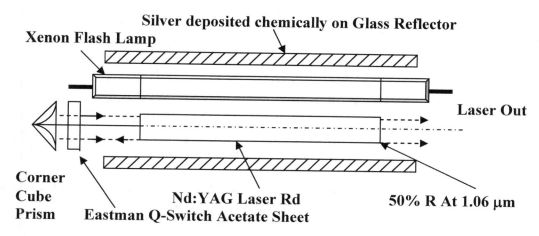

Figure-8: Nd:YAG Laser (for Hand-held Range Finder)

64

3.3.4 Erbium Doped Glass Laser

Erbium (Er) is a rare earth element with 12 electrons in its 4f unfilled inner shell with electronic configuration as

$$4d^{10}, 4f^{12}, 5s^2, 5p^6, ..., 6s^2$$

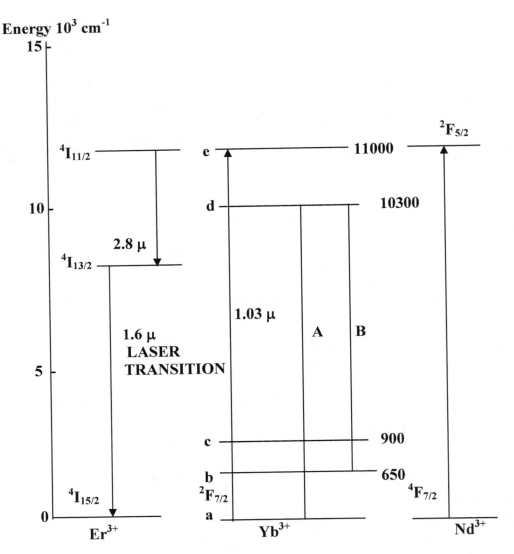

Figure-9: Energy Level Diagram of Er³⁺ with Yb³⁺ and Nd³⁺ as Sensitizer in Glass

Laser oscillations in this material are observed in wavelength region from 1.53 to 1.66 μm. This material has importance in the range finders because at this wavelength eye is less subject to retinal damage. Er is doped in silicate or phosphate glass with sensitizer such as Nd, Cr or Ytterbium (Yb) to absorb more pump light from flash lamp, as Er is relatively transparent to xenon flash lamp output. Er doped silicate or phosphate glass generates light at 1.54 μm. The principal absorption band of Er^{3+} ion is at 1 μm. Xenon flash lamp has the spectral output that matches the pumping requirement of Yb^{3+}. The energy level scheme of Er^{3+} ion as shown in the figure-9 is from metastable level $^4I_{13/2}$ to ground state $^4I_{15/2}$ for laser transition at 1.54 μm. The optimum Er concentration in glass is between 0.2 to 0.5 % by weight of Er_2O_3. However, the Er^{3+} ion concentration must be sufficiently high to produce high energy-transfer efficiency from Yb^{3+} ions to Er^{3+} ions.

3.3.5 Nd:YAG-Methane (CH₄) Raman Laser:

The Nd:YAG-Methane Raman laser operates at 1.54 μm is also being used in eye safe laser range finders. Nd:YAG laser beam having intensity greater than 10^6 watts per cm^2 is passed through a high pressure methane cell with pressure between 50 to 80 atmosphere. Interaction of photons with methane due to resonant Raman effect produce stimulated emission at 1.54 μm from 1.064 μm. These photons are amplified by excited methane molecules as shown in energy level diagram of methane in figure-10. Due to initial interaction between unexcited methane molecules and photons at 1.064μm some laser energy is lost in strokes lines, but maximum 65% energy conversion efficiency can be achieved depending upon pressure of methane gas, cell length and intensity of Nd:YAG laser. In this laser beam collimation and pulse duration is reduced. Only advantage is in modifying existing Nd:YAG laser range finder into eye safe one without changing laser material.

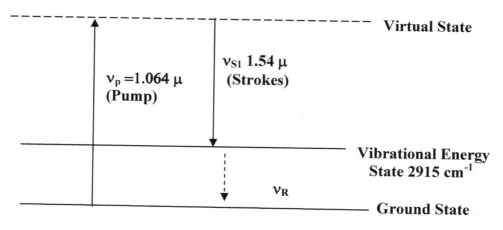

Figure-10: Raman Process Showing First Stroke Shift

3.3.6 Optical Parametric Oscillator (OPO): Now days OPO are widely used for ranging in mid infrared wavelength. These lasers does not require elaborate cooling techniques, can give wide tuning range with high power and narrow pulse width in mid infrared, only by changing orientation of non-linear crystal any light from 1 to 6 μm can be generated. Diode pumped Nd:YAG laser is very good pump source for non-linear crystals. Generally periodically poled lithium niobate (PPLN) is used. This source can be used in differential absorption (DIAL) system; natural target at few kilometers can be used for measurements of species in atmosphere, since they produce narrow line width, high peak power pulses. Output of high power Nd:YAG laser at 1064 nm is converted into a pair of lower frequency, signal and idler photon while conversing the total energy and momentum of pump photon.

Tunability of the signal-idler pair is usually achieved either by changing the crystal birefringence through its temperature dependence or by angular dependence of the extra ordinary index of crystal. The practical optical parametric oscillator (OPO) consist of one or more non-linear crystal enclosed in an optical cavity resonant at either or both the signal or idler wavelength and pumped by either the fundamental or a harmonic frequency of Nd:YAG laser. The output of the OPO can be amplified by phase matched mixing of the signal or idler photons with the pump photons in similar crystal. Experimental set up for optical parametric oscillator (OPO) with angle tuning and temperature tuning is illustrated in figure-11. A new generation of multi-watt cw singly resonant OPO based on diode pumped solid state laser (DPSS), periodically poled lithium niobate (PPLN) is placed within a four-mirror ring cavity is shown in figure-12. The high non-linearity of PPLN decreases pump requirements of singly resonant devices to modest levels, i.e. threshold of 2 to 4 watts, efficiencies achieved exceeding 90%, and idler wavelengths can vary from 2.3 to 4.7 μm. Subjecting the crystal to alternating electric fields during growth makes periodically poled crystals. Poled regions with spacing of the order of 10 μm or less can be created. Recently a microchip OPO as shown in figure-13

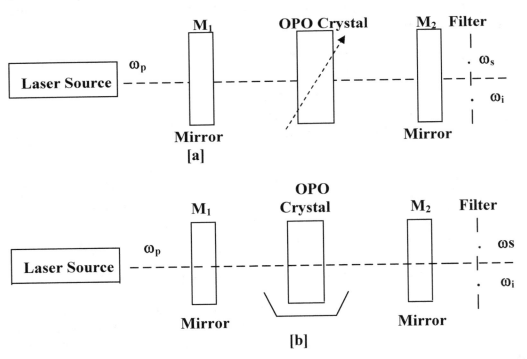

Figure-11: Experimental Set Up for Optical Parametric Oscillator (OPO) with [a] Angle Tuning and [b] Temperature Tuning.

have been developed which operates in CW mode up to 6 μm wavelength with diode pumped Yb^{3+} periodically poled titanium niobate crystal. MgO is doped to prevent refractive damage. [19,20].

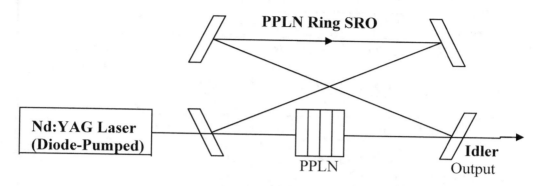

Figure-12 Continuous-wave (CW) Single Resonant OPO (SRO) Composed of Four Mirrors Ring. (Permission and Courtesy of Laser Focus World)

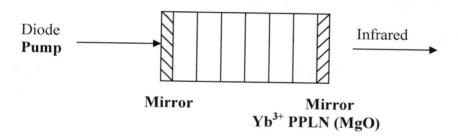

Figure-13: PPLN Microchip OPO with Very Low Power – 0.5 mW. (Permission and Courtesy of Laser Focus World)

Nd:YAG Pumped KTP OPO Nd:YAG laser Pumping KTP crystal to generate laser output at 1576 nm has been developed at IRDE as eye safe laser source for target designator and for ranging aerial targets. It consists of 5 mm X 50 mm Nd:YAG laser rod pumped from one side by four proximally placed quasi continuous wave (QCW) 800 W output at 808 nm laser diode arrays operated for 200 μs at pulse repetition rate of 20 pulses per second as shown in figure-14. A crossed porro resonator with polarization coupled with electro-optic Q-Switching is used. With this input power a laser pulse of energy 85 mJ in 17 ns pulse duration is generated at 1064 nm. This laser pulse output is down converted to 1567 nm using non-critically phase matched (NCPM) monolithic KTP with conversion efficiency of 27%. This laser source is used in range finder which is eye safe, with better beam propagation in adverse weather conditions as compared to Nd:YAG laser source at 1064 nm.

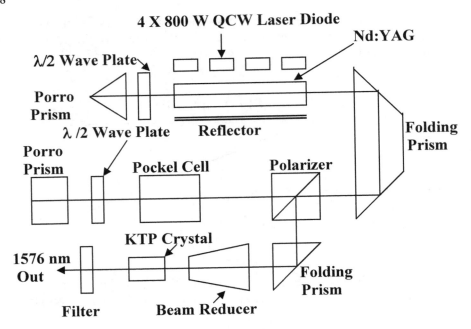

Figure-14: Optical Parametric Oscillator with laser output at 1567 nm

3.4 GAS LASER

The following gas laser sources are being used in range finders and distance measuring equipment.

3.4.1 Helium-Neon (He-Ne) Laser

The He-Ne laser is first continuous wave (cw) laser demonstrated in 1962 by White and Rigdon [13]. This laser produces highly collimated beam at 0.6328 μm and find application for distance-measuring equipment to measure accurately within millimeters (mm) range of target with retro-reflector for survey purpose. The laser was intensity modulated with different sinusoidal frequency and by comparing phase of reflected light, range of target fitted with retro-reflector at a distance of few kilometers can be measured within mm accuracy. The detector used in receiver of equipment was photo multiplier with S-20 response.

The He-Ne laser consist of a long discharge tube in which He is filled at pressure of 1 mm of Mercury (Hg) and Ne at 0.1 mm of Hg. Laser action in He-Ne gas takes place in visible and infrared wave length of light. The wavelength can be chosen using wavelength selective mirrors. The laser action at 0.6328 μm is produced from meta stable state $3s_2$ to terminal level $2p_4$ of Ne gas as shown in energy level figure-15. A 1.1523 μm laser line is produced from $2s_2$ meta stable state to same terminal level $2p_4$ of Neon. Excitation of He atoms in gas discharge is used as resonant collision to raise unexcited Ne atoms to 2s and 3s meta stable state. Since 2^3s level of He has same energy to 2s level of Neon, 2^1s level of He has same energy as 3s level of Neon. Excitation of Neon atom

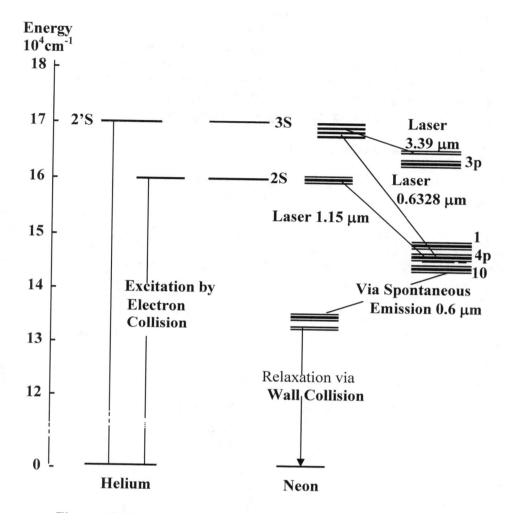

Figure-15: Energy Level Diagram of He-Ne Laser.

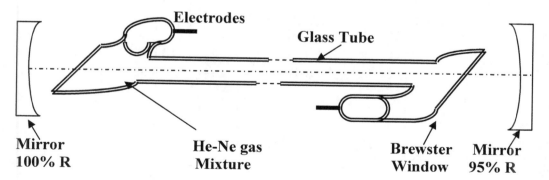

Figure-16: A DC Excited He-Ne Gas Laser

70

by He excited atom is more efficient than direct excitation of Neon atoms. The output power of He-Ne laser is generally 0.5 to 5 mW range. Both the 0.6328 and 3.39 μm transitions start from the same 3s level. The 3.39 μm transition has a higher gain and thus the laser tends to lase at this frequency unless precautions are taken. This might to ensure that the cavity mirrors have small reflection coefficient at 3.39 μm. A direct current excited He-Ne laser [14] developed at IRDE is shown in figure –16. This laser at 3.39 μm is used for measurement of methane [28] in atmosphere.

3.4.2 Carbon dioxide (CO_2) Laser

The CO_2 laser is a molecular laser where the vibrational levels of one carbon atom and two oxygen atoms, bonded by chemical are used. Except for semiconductor lasers, it has the highest overall efficiency of 30%. It produces millions of watt cw wave power and even higher amount in pulse mode. Since CO_2 is a tri atomic linear molecule, it has three vibrational modes, v_1, v_2, v_3 as shown in figure-18 v_1 is a symmetrical stretching vibration (Expansion and contraction), v_2 is a doubly degenerate bending vibration, and v_3 is an anti symmetrical stretching vibration. The ground level (0,0,0) and

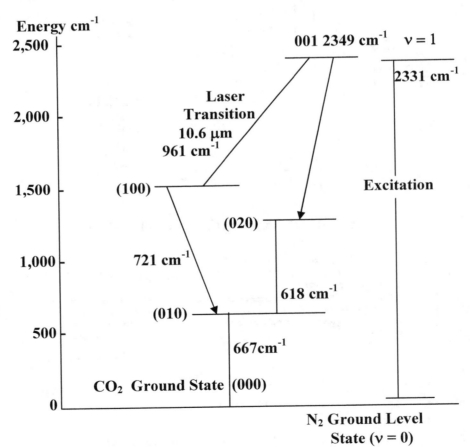

Figure-17: Energy Level Diagram of CO_2

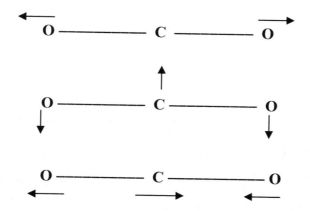

Figure-18: Three Modes of Vibration in CO$_2$

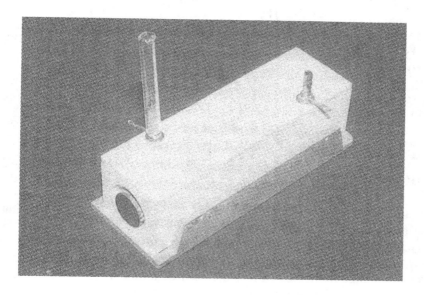

Photo-1: TEA CO$_2$ Plasma Tube

excited energy levels are 001 and 010 shown in energy level diagram (figure-17). The CO$_2$ laser uses a mixture of He, N$_2$ and CO$_2$. The excited vibration state of N$_2$, which is meta stable state, has energy close to vibration level v_3 of CO$_2$. Thus from this meta stable state of N$_2$ resonant energy to unexcited CO$_2$ molecules takes place by resonant energy transfer. The role of He is to depopulate the lower terminal levels (I, 0, 0) and (0, 2, 0) for efficient laser transition to these terminal levels. The normal cw CO$_2$ laser is excited by direct current with a gas pressure of about 1 to 5 Torr, but pulse discharge through transfer excitation at atmospheric pressure i.e. TEA laser gives sharp pulses with peak power of 0.5 to 100 MW in 50 to 100 nano second duration pulses. This laser is used for laser ranging under poor visible conditions. A compact TEA CO$_2$ laser source [15] developed at IRDE is shown in figure-19 and in photo-1.

72

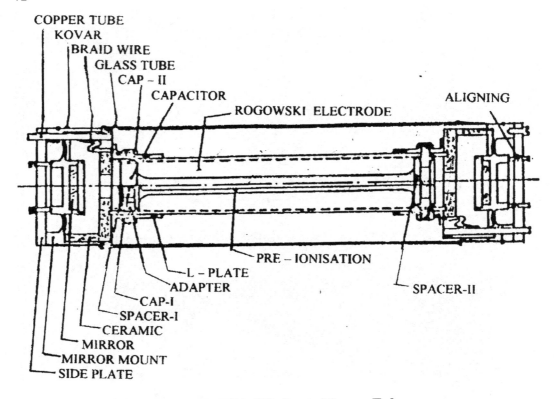

COPPER TUBE
KOVAR
BRAID WIRE
GLASS TUBE
CAP – II
CAPACITOR
ROGOWSKI ELECTRODE
ALIGNING
PRE – IONISATION
L – PLATE
ADAPTER
CAP-I
SPACER-I
CERAMIC
MIRROR
MIRROR MOUNT
SIDE PLATE
SPACER-II

Figure-19: TEA CO$_2$ Laser Plasma Tube

3.5 SEMICONDUCTOR

A semiconductor [16] is a material with electrical conductivity more than good insulator, yet much less than metal. In this type of material, electrons are held between adjacent atoms by covalent bonds. Occasionally, an individual electron of a covalent bond acquires sufficient energy to break the bond and become free. These electrons move in the crystal in a random way like the molecules of a gas. When an electric field is applied, there is a steady drift beside this random motion, towards positive electrode that represent a flow of current carried by electrons. When the electron breaks away from covalent bond, the empty place left behind is known as hole. When a free electron moves randomly in the crystal, there is possibility that it will encounter a hole. This does not happen very frequently, but it happens, the electron reestablishes the missing covalent bond. With this recombination free electron is eliminated. The combination of electrons and holes is proportional to the product of concentration of electrons and holes. The lifetime of an individual carrier is limited; typically it ranges from 1 μ sec. to 1 millisecond according to circumstances. The concentration of electrons denoted by n and holes denoted by p in a pure semiconductor build up to a level such that the rate of recombination of holes and electrons equals to their rate of production. With increase in temperature T, concentration of holes and electron increases. In pure semiconductor called intrinsic, the concentration of electrons n_i is given by relation

$$n_i = A \ T^{3/2} \ e^{-eE/2kT}$$

..........[1]

Where E is energy required to break covalent bond of semiconductor and A depends on nature of semiconductor.

If μ_n and μ_p is mobility of free electrons and holes in a semiconductor, the conductivity is given by

$$\sigma_I = e \ n_i \ (\mu_p + \mu_n)$$

..........[2]

Impurity in Semiconductor: A very small amount of impurity in semiconductor alters concentration of electrons and holes carrier to a greater extent.

3.5.1 Donor or n-Type Semiconductor:

Atoms such as arsenic (As), antimony (Sb) or phosphorous (P) that is pentavalent is added to tetravalent semiconductor, four valency electrons of impurity atoms are held by adjacent atoms of semiconductor by covalent bonds, while fifth valency electron of impurity atom which is unbound can be removed with little force. This electron moves randomly through the crystal in semiconductor. When an electric field is applied, there is steady drift of this free electron towards positive electrode that represent a flow of current. When the impurity atom loses an electron, it becomes a positively charged ion. This ion is immobile, however it is held by covalent bonds to the adjacent atom, and so cannot contribute to the conduction of electricity. Impurity atoms that that contribute free electrons are called donor, because they denote free electrons. Semiconductor containing this type of atoms is called n-type semiconductor. The free electrons supplied in this way are called excess electrons.

3.5.2 Acceptor or p-Type Semiconductor:

If trivalent impurity, like aluminum (Al), boron (B), Gallium (Ga) or indium (In) is added to a tetravalent semiconductor, the impurity atoms form covalent bond with three neighboring atoms of semiconductor, while fourth atom of semiconductor, is short of one electron to form covalent bond, therefore it has tendency to capture free moving electron, or capture electron from neighboring atom, it then become a mobile positive ion. The adjacent vacancy of electron of the neighboring atom moves in a random way due to thermal effects, and when an electric field is applied, it tends to drift towards negative electrode and this contribute to flow of current. Impurity atoms that contribute holes in this manner are called acceptor. Semiconductors containing acceptor are called p-type semiconductors. Holes created in this way are called excess holes.

It is found that n or p-type of semiconductors, product of electrons and holes is equal to number of intrinsic electrons i.e. $n_i^2 = n \cdot p$ and current due to these electrons and holes is given by

$$\sigma_n = e \ N_d \ \mu_n$$

.........[3]

74

$$\sigma_p = e\, N_a\, \mu_p \qquad\qquad\qquad \dots\dots\dots[4]$$

Where N_a and N_d are concentration of acceptor and donor atoms.

Firmi-Dirac Distribution: The density of electron with energy E is given by Firmi-Dirac distribution function f

$$f = [1 + e^{(E-F)/kT}]^{-1} \qquad\qquad \dots\dots\dots\dots[5]$$

Where F- Fermi level

T- Absolute temperature, above 0°K some electrons are always present in the conduction band. The energy of electron is actual a function of its moment p.

3.6 SEMICONDUCTOR LASER

The semiconductor lasers are commonly known as diode laser. But their different forms are p-n junction, single or double hetro-structure (SH or DH) laser, quantum-well laser depending on their structure. The Laser emission was first observed simultaneously in forward-biased, highly doped (degenerate) Ga-As p-n junction cooled to 77°K by Hall et al [17] and Nathan et al [18]. The radiation, which originates in the vicinity of the junction, is due to transition of injected electrons and holes between the low lying levels of the conduction band (donor) and the upper most level of the valence band (acceptors). The frequency of the emitted radiation i.e. 0.842 μm corresponds to

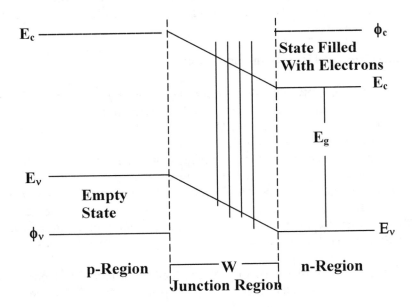

Figure-20- Forward Bias p-n Junction of Heavily Doped Semiconductor Laser with Recombination Occurring in Junction Region.

band gap energy. The laser was operated in pulse mode with pulse width unto 20 μ second. The laser action is illustrated in figure-20. These lasers were used only in late eighties in range finders and distance measuring equipment for survey purpose replacing He-Ne lasers with the development of reliable high pulse rate hetro-structure Ga-As lasers operating at ambient field temperature conditions. Using microprocessor based signal detection and time averaging techniques, range from diffuse reflecting targets could be measured up to few hundreds meter with range accuracy within a cm.

3.6.1 Hetero Structure Semiconductor Laser

Hetero structure semiconductor laser is basically a p-i-n diode laser. When forward biased, electrons in the conduction band and holes in the valence band are injected into the intrinsic region called active region. The electrons and the holes accumulate in the active region, where they are induced to recombine under the action of laser field confined in the same region [21].

It consists of GaAs with $Al_x Ga_{1-x}$ As (abbreviated to AlGaAs) hetero structure as shown in figure-21. Band gap energy of GaAlAs is greater than that of GaAs, with smaller refractive index. Therefore in the hetero structure of GaAs and AlGaAs, the light is concentrated in GaAs, whose refractive index is greater, also injected carrier electrons and holes are concentrated in GaAs because the band gap is narrower. Thus if active layer of GaAs is sandwiched between p-type of AlGaAs and n-type AlGaAs, both light and carriers are confined in the active layer. This leads to strong interaction with reduction in light losses. This type of junctions is called double heterostructures (DH). Operation in this type of laser is possible in pulse and cw mode in normal ambient temperature with appropriate cooling. The length of semiconductor is normally 100 to 500 μm, the stripe is 2 - 20 μm and the thickness of active layer is 0.1 to 2 μm and cw power of I - 100 mW with peak power of 1-10 W in pulse mode, with pulse width of the order of a pico-second can be easily obtained.

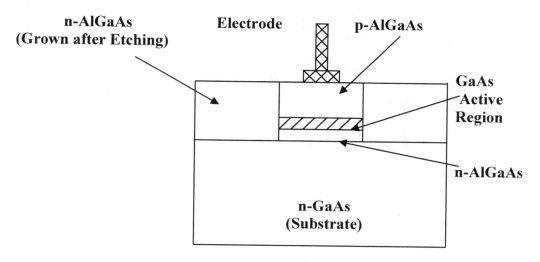

Figure-21- Cross-Section of a Buried Double Hetero Structure Laser

76

3.6.2 Quantum Cascade Lasers (QC): Federico Capasso and Jerome Faist at Bell Labs invented the QC laser in 1994. These lasers are inherently high power devices due to the cascading scheme and consist of stacks of alternate thin layers (order of nanometers) of semiconductor material like GaAs and AlGaAs or InGaAs and AlInAs. Since layers are so thin that two energy levels of electrons in conduction band are created by quantum confinement. Energy difference in these levels depends upon thickness of these layers. There can be 25 to 75 stacks of such layers in a QC laser. Unlike in normal semiconductors with electrons and holes, in QC layers when electric current is passed, electron will jump from higher energy level to low energy level in conduction band emitting a photon. Depending upon number of stacks of these layers, same election emits number of photons while crossing layers. Number of photons or quantum yields from single electron can be 50% of number of layers. At present QC lasers are tailored to operate in mid infrared region for detection of methane or nitrous oxide and other green house gases in atmosphere. Recent advance are for development of wide tunable single mode distributed feedback (DFB) and bi-directional for fine spectroscopy and differential lidar. Output power of these lasers is ten times that of lead salt quantum well lasers. These lasers operate at room temperature in 4 to 5 micron wavelength range in pulse mode and operate at higher wavelength, cooling them with thermoelectric coolers. Figure-22 shows conduction band profile of two active regions connected by an injector under positive bias relevant wave functions and the first mini-band of the injector. The laser transition is indicated by wavy arrows structure with incorporated AlAs blocking barriers. This layer sequence of one active region and injector, in nanometers from left to right starting from the injector barriers is **5.0**, 1.0, **1.5**, 2.0, 0.7 (InAs), 2.0, **2.2**, 4.1, **0.9, 0.7 (AlAs)** ,**0.9**, 2.5, **2.3**,2.3, 2.2, 2.0,**2.0**, *2.0, 2.3, 1.9*, **2.8**,1.9. The $Al_{0.48}In_{0.52}As$ barriers ($Ga_{0.47}In_{0.53}As$ quantum well) are type set in bolt (roman), lasers in italic are Si doped to n $= 2x10^{17}$ cm^{-3} and the underline layers serve as the exit barrier [23, 24].

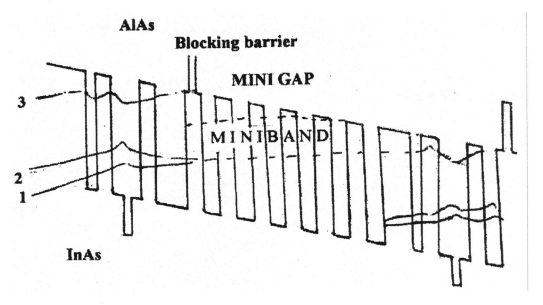

Figure-22: Schematic of conduction band profile of active region of Quantum Cascade Laser. (Courtesy of Applied Physics Letters)

3.6.3 Quantum Well Lasers: These lasers work in visible and near infrared region with better stability, tune ability and efficiency. These lasers are based on alloys of group III and V elements and emit light in the spectral range of 0.4 to 2.5 micrometer. Major advantage with these lasers is they can operate at room temperature or require little cooling for better efficiency. They find wide use in CD players, printers and fiber communication. With high sensitive detectors available, these lasers also are used for detection of carbon dioxide, carbon monoxide, nitrogen dioxide, ammonia and methane using differential spectroscopy in near infrared on weaker absorption lines of these species with better accuracy and low concentration level. [21, 25]

3.7 CONCLUSION

Although, better and efficient crystal materials have been developed, but due to ruggedness, good heat conduction, better beam quality and stability, diode pumped Nd:YAG will be preffered in laser ranging instrument for high repetition rate laser range finder underwater ranging with second harmonics, compact rangefinders or mid infrared rangefinders (Nd:YAG Pumped OPO) with thermal sights working under poor visible conditions.

SUMMARY

1. Ruby operating in pulse mode at 0.6943 μm is the first laser source to be used successfully in laser range finders.

2. He-Ne laser in cw mode of operation with sinusoidal intensity modulation has been used for measuring distance of targets with retro-reflector i.e. co-operative target for survey purpose.

3. With the development of silicon avalanche detector (Si APD), Nd:Glass and Nd:YAG laser sources are used in range finders.

4. Now gallium aluminum arsenide (GaAlAs) laser source in high pulse repetition rate is being used in distance measuring equipment.

5. Now with the development of Indium Gallium Phosphate avalanche detector, Er Glass source is commonly used for eye safe range finders.

6. TEA CO_2 Pulsed source is used in laser range finder using thermal sight due to better propagation under hazy conditions.

7. OPO laser sources using efficient non linear crystal, pumped by Nd:YAG lasers is present state of art to be used in eye safe and mid infrared range finders giving better capabilities for ranging under poor visible conditions using thermal sight for military targets.

APPENDIX-A

Solidstate Laser [8, 27, 29]

Characteristics	Ruby	Nd:YAG	Alexandrite	Ti:Sapphire
Structure	Hexagonal	Cubic	Orthorhombic	Hexagonal
Hardness	9.0	8.2	8.5	9.1
Thermal conductivity (W/cm-K) at 300 K	0.42	0.14	0.23	0.42
Upper state lifetime	3.0 ms	230 μs	260 μs	3.2 μs
Cross-section cm^2	2.5 x10^{-20}	6.5 x 10^{-19}	1 x 10^{-20}	4.1 x 10^{-19}
Line width (A^0)	5.3 (300K)	4.5 (300K)	1000	2300

Nd Doped Glasses [12]

Characteristics	*Silicate (Q-245)	*Phosphate (Q-100)
Peak Wavelength [nm]	1062	1054
Cross Section [x 10^{20}]	2.9	4.4
Fluorescent Lifetime [μs]	340	190
Line width [nm]	27.7	21.2
Density [gm/cc]	2.55	3.204
Refractive Index	1.568	1.562
Thermal Conductivity [W/m]	1.30	0.82
Knoop Hardness	600	558

*Manufacturer M/S Kigre, Inc USA

(i) **Ti: Sapphire Laser:** Ti:Al$_2$O$_3$ laser has an exceptionally wide tuning range and a large gain cross-section. Large high quality crystals are commercially available and consist of sapphire doped with 0.1% of Ti^{+3} by weight. Its vibronic terminal band allows tunable laser output between 670 to 1070 nm with peak gain around 800 nm. Ti:Sapphire laser has absorption band in visible with peak absorption at 490 nm and has been pumped with a number of source like second harmonic Nd:YAG and Nd:YLF laser, copper vapor laser and argon laser beside xenon flash lamp. Due to small lifetime 2.3 μs of metastable state, pulsed 2nd harmonic Nd:YAG or Nd:YLF is preferred for pumping this material for high repetition rate for range gated laser range finders or its third harmonic is used for detection and measurement of ozone in atmosphere.

(ii) **Alexandride Laser (Cr^{+3}:BeAl$_2$O$_4$):** Alexandride laser is a solid-state laser in which Chromium ions (Cr^{+3}); at the amount of 0.01-0.4%, are embedded in BeAl$_2$O$_4$ crystal. It has energy level structure similar to the energy level structure of Ruby laser. Alexandride laser was operated for the first time as a three level laser in 1973 at a wavelength of 680 nm. A few years later, it was found that at longer wavelengths the Alexandrite laser could be operated as a four level laser, which can be tuned over a 720-800 nm range of wavelength. Since its terminal level can be vibrational energy band as shown in energy level diagram (figure-1A). The absorption bands of Alexandride laser

80

are very similar to ruby material with peaks at 410 and 590 nm. Gain cross-section for laser action increase with temperature. This results in improved performance at elevated temperature. This material is widely used for range-gated illumination under adverse weather conditions [27].

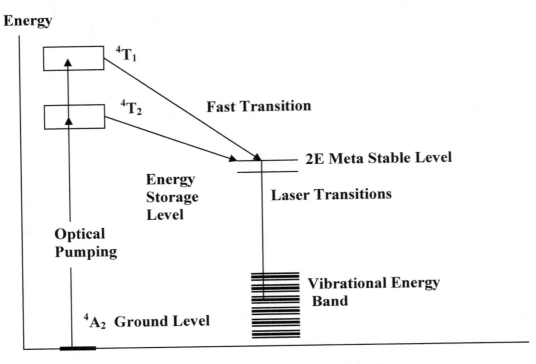

Figure-1A: Energy Level Diagram of Alexandrite Laser.

REFERENCES

1. Maiman,T.H. "Stimulated Optical Radiation in Ruby Masers", Nature, Vol. 187, pp.493 - 494, August 1960.

2. Hellwarth, R.W., Advances in Quantum Electronics, Columbia University Press, New York, (1961).

3. Benson, R.C. and Mirarchi, M.R., the Spinning Reflector Techniques for Ruby Laser Pulse Control", IEEE Transactions on Military Electronics, Vol. 8, No.1,January 1964, pp. 13-21.

4. Yariv, A. , and J. P. Gordon, " The Laser, " Proc. IEEE, Vol.51, No.1, January 1963, pp. 4 -29.

5. Steele, L. Earl, "Optical Laser in Electronics", John Wiley and Sons Inc., New York, (1968).

6. Lengyel, A. , Bela, "Lasers", (2nd Edition) Wiley-Interscience, New York (1971).

7. Patek,K., "Glass Lasers", Butterworth & Co (Publisher) Ltd. London, (1970).

8. Koechner, Walter, "Solid State Laser Engineering," Springer Series in Optical Sciences, Springer-Verlag, Berlin (1992).

9. Snitzer,E., "Optical Maser Action of Nd^{+3} in Barium Crown Glass", Phy. Rev. Lett., Vol. 7, December 1961, pp. 444-446.

10. Mansharamani,N., Rampal,V.V., and Srivastava,K.P., "Laser Action of C.G.& C.R.I.Nd:Glass – A comparative Study", Journal of Instrument Society of India, Vol. 6, No.2, 1976.

11. Kuhn,K.J., "Laser Engineering", Prentice Hall, NJ 07632, USA (1998).

12. M/S Kigre Inc., USA, Product Catalogue.

13. White,A.D. and Rigden,J.R., "Continuous Gas Maser Operation in the Visible", Proc. IRE (Correspondence), Vol. 50, July 1962, pp. 1697.

14. Singh,R.N., Juyal, D.P., Bhargava,J.S., "Design of a Continuous He-Ne Laser Operating at 6328 A^0," Presented at Symposium on Science and Technology of Infrared and Laser held on 2-3 February 1970 at Defence Science Laboratory, Delhi.

15. Juyal,D.P., et.al., "A Compact Pulsed TEA CO_2 Laser", Journal of Institute of Electronics & Telecommunication Engineers", Vol. 33, No.6, 1987, pp. 187-189.

16. Terman,F.E., "Electronics and Radio Engineering", 4th Edition, McGraw-Hill Book Company, Inc., New York, (1955).

17. Hall,R.N., et. al., "Coherent Light Emission from GaAs Junction", Phys. Rev. Lett., Vol. 9, November 1, 1962, pp. 366-367.

18. Nathan,M.I., et. al., Stimulated Emission of Radiation from GaAs, p-n Junction, App.Phys. Lett., Vol. 1, November 1963, pp. 62-64.

19. Capmany,J., et.al. "Micrchip OPO's Operate in the infrared", Laser Focus World, Vol.37, No.6 June 2001, pp. 143-148.

20. Guang,S.H., and Lice,S.H., "Physics of Nonlinear Optics", Publisher-World Scientific, Singapore, NJ, London, Hong Kong (1999).

21. Kapoon,E., "Semiconductor Laser-I Fundamentals", Academic Press, London, (1999).

22. Chih-Wie Hsu, Yang,C.C., "Broadband Infrared Generation with Non-Collinear Optical Parametric Processes on Periodically Poled LiNbO$_3$", Optics Letters, Vol. 26, No. 18, September 2001, pp. 1412-1414.

23. Mann,Ch, et al., "Quantum Cascade Lasers for the Mid-Infrared Spectral Range: Devices and Applications", Advances in Solid State Physics, Vol. 43 by B.Kramer (Ed.), pp. 351-368.

24. Yang,Q.K, et al., "Improvement of λ =5 μm Quantum Cascade Lasers by Blocking Barriers in the Active Region", Appl. Phys. Lett., Vol. 80, 2002, pp. 2048-2050.

25. Christopher L. Felix, et. al. "Mid-Infrared W. Quantum-Well Lasers for Non-Cryogenic Continuous-Wave Operation", Appl. Optics (USA), Vol. 40, No. 6, February 2001, pp.806-811.

26. Mansharamani, N., "Greenhouse-Effect-Lidar Techniques", ISBN:81-7525-789-X Sita Publisher, Dehradun, India, (2006).

27. Sam, R.C., "Alexandrite Lasers", Handbook of Solid State Lasers, Ed. Peter,KChoe, Vol.18, pp.444-449 , Optical Engineering, Pub. Marcel Dekker Inc, USA, (1989).

28. Mc Manus,J.B., Kebabian,P.L. and Kolb,C.E., "Atmospheric Methane Measurement Instrument Using a Zeeman Split He-Ne Laser", Appl. Optics (USA), Vol. 28, No. 23, 1 December 1989, pp.5016-5023.

29. M/S Airton, USA, Product Catalog.

CHAPTER-4

OPICAL RESONATOR

4.1 INTRODUCTION

The optical resonator or optical laser cavity consist of two spaced plane parallel or co-axis spherical mirrors in which light can be made to move to and fro through laser material kept in the resonator. For reflection from mirrors light should satisfy phase conditions for reflection to take place i.e. spacing L between the mirror should be equal to integer multiple of half wavelength of light. The frequency difference Δv between two adjacent frequencies with difference equal to one integer of half wavelength in free space length L in a resonator is c/2L. Therefore in a resonator there are number of discrete frequencies as illustrated in figure-1, which depend on gain and line width of laser materials. These discrete frequencies are known as longitudinal modes of resonator.

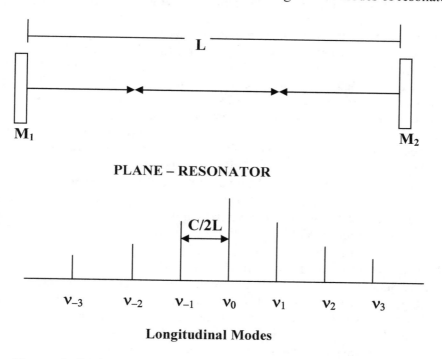

Figure-1: Optical Resonator and its Longitudinal Modes

Beside this, due to finite width of resonator, the electric field vectors direction reverses in transverse direction inside and outside resonator depending on gain and width of laser material. The numbers of field reversal in horizontal and vertical direction are known as transverse modes. The modes in a resonator with rectangular cross section are denoted by letter TE_{mnq}, where m and n represent number of electric field reversal in x and y direction respectively, q represent number of half wavelength in length L of resonator. TE_{plq} represents the modes in a resonator with circular mirrors, where p, l and q are numbers of electrical nodes in r, θ, and z direction. Depending on the configuration and geometry of resonator, resonators are classified as stable and unstable i.e. in stable resonator, bundles of light rays are periodically refocused as it travels back and forth between the two mirrors. In unstable resonator, light is dispersed more and more as it travels to and fro between mirrors. In this chapter, beside modes in resonator, various types of resonators are described that are mainly used for rangefinders for better beam quality, stability and low divergence.

4.2 HISTORICAL BACKGROUND

Before the first ruby and He-Ne lasers were demonstrated, Basov and Prokhorov, Dicke, and Schawlow and Townes proposed Fabry-Perot type cavity for generation of stimulated emission. Schawlow and Towns research paper in which they discussed crucial point in laser design and discussed their choice of a Fabry-Perot cavity was received on 26th August 1958 in Physical Review and published in December 1958 [1]. Schawlow, a student under Prof. Malcolm F.Crawfold in Toronto used Fabry-Perot interferometer to carry out research on the hyperfine structure in atomic spectra. The approximate theory of the plane-mirror resonator, as developed by Schawlow and Town in above paper, requires a more detail analysis for diffraction losses. The first satisfactory approach was due to Fox and Li [2] in 1960 that investigated the effect of diffraction on the electromagnetic field in a Fabry-Perot interferometer in free space. Their approach was to consider a propagating wave reflected back and forth by two parallel, plane mirror, is equivalent to the case of a transmission medium comprising of a series of collinear, identical aperture cut into parallel and equally spaced black partitions of infinite extent. The self-reproducing distribution of phase and relative amplitude once the aperture that they found may be regarded as a proper mode of the cavity. A parameter which was useful in calculation was the quantity $N = a^2/\lambda L$ as Fresnel number for circular mirrors of radius a with spacing L. When $L << a^2/\lambda$, the center of one mirror is the near field of other regarded as aperture, and the field in the central region of the second mirror may be calculated from the field on the first by means of geometrical optics. When $L > a^2/\lambda$, the Fresnel zones aperture and geometrical optics were not adequate for the calculations. After Towns publication in 1958, Bell Laboratory filed a patent application in the United States on 6th April 1959. Before that Town wanted from Gorden Gould who was working on a private project at TRW Inc, on the very bright thallium lamp that Gould [4] has used when he was research assistant at Columbia radiation Laboratory, to produce optically excited state in beam of thallium. Gould, on 16th November 1957, got his laboratory notebook that contain the work entitled, "Some rough calculation on the feasibility of laser light amplification by stimulated emission of radiation" from public notary. After the demonstration of ruby laser by Maiman in 1960, there was a legal battle between Town and Gould. However, on publication in Phys Review, C.H.Towen was awarded Nobel Prize in 1964 and A.L.Schawlow in 1981 for a related subject: Laser Spectroscopy. After 17 years of legal battle Gould finally received patent on an optically pumped laser amplifier on 11th October 1977.

4.3 RESONATOR MODES

There are two types of modes for light to form standing waves in a resonator

4.3.1 Longitudinal Modes

Let us consider a Fabry- Perot resonator with L as free space length between the mirrors. The light waves to form standing waves between in a resonator, the free space length between the mirror should be equal to multiple integer of half wavelength of light. Following relation gives the frequency spacing between the adjacent longitudinal modes in a resonator

$$\Delta v = c/2L \qquad \qquad \text{.......[1]}$$

Where c is velocity of light in space.

The frequency spacing between modes increases with decrease in resonator length.

4.3.2 Transverse Modes

If TE_{mn} is transverse mode in a resonator with rectangular mirrors, n and m denotes number of nodes in x and y direction, then the intensity I_{mn} and electric field E distribution as derived [2,5,6,7,8] is given as below

$$I_{mn}(x,y,z) = I_0 \left[H_m \left\{ \frac{x(2)^{1/2}}{w(z)} \right\} exp. \left\{ \frac{-x^2}{w^2(z)} \right\} \right]^2 \times \left[H_n \left\{ \frac{y(2)^{1/2}}{w(z)} \right\} exp. \left\{ \frac{-y^2}{w^2(z)} \right\} \right]^2 \quad ...[2]$$

The $H_m(s)$ is the mth order Hermite polynomial that is $H_0(s) = 1$, $H_1(s) = 2s$, $H_2(s) = 4s^2-2$

w(z) is the spot size in which transverse intensity decrease to $1/e^2$ of the peak intensity of the lowest-order mode. The m, n values can be determined by counting the number of dark bars crossing the intensity pattern in the x and y directions.

$$E(x,y) = E_0 \, H_m[(x/w).(2)^{1/2}] \, H_n[(y/w).(2)^{1/2}] \, exp[-(x^2 + y^2)w^2] \qquad \text{......[3]}$$

E_0 = constant amplitude.
$H_n(x)$ = Hermite polynomial of nth order
w(z) = spot size at distance z.

If TEM_{pl} are transverse electric modes of resonator with circular mirror, then relation below gives the normalized radial intensity distribution of light beam.

$$I_{pl}(r,\phi,z) = I_0 \rho^l [L_p^l \, \rho]^2 (cos^2 l\phi) \, exp(- \rho) \qquad \text{............[4]}$$

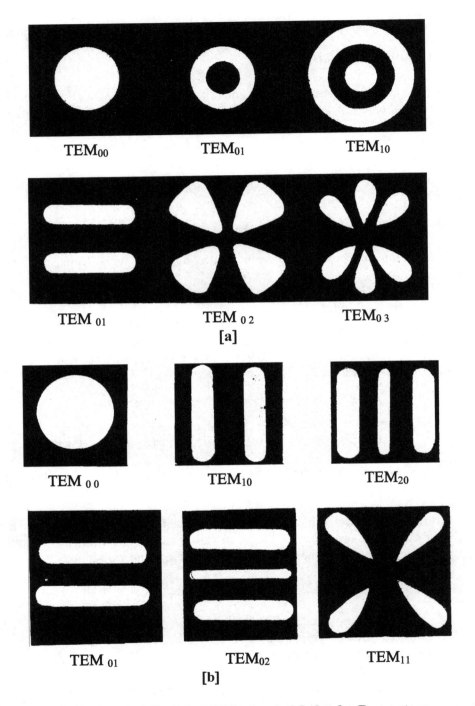

Figure-2 Intensity Distribution of Transverse Modes for Resonators With [a] Circular Mirrors [b] Rectangular Mirrors

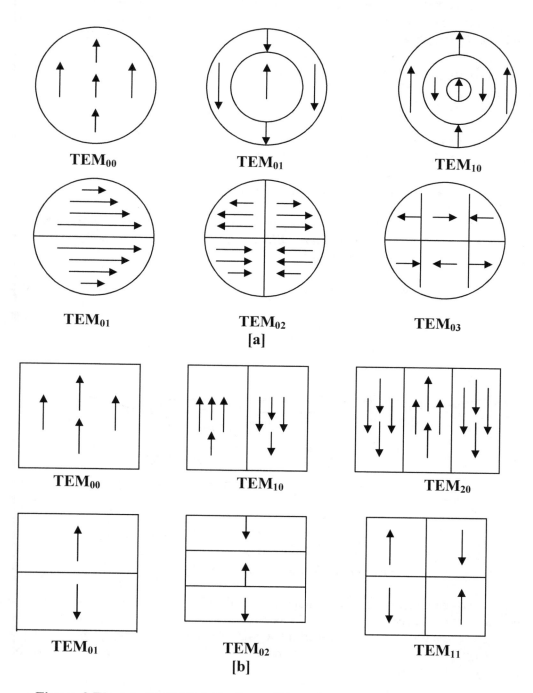

Figure-3 Electric Field Distribution of Transfer Modes for Resonator with [a] Circular Mirrors [b] Rectangular Mirrors

88

Where z is the direction of beam propagation, and r, ϕ are the polar coordinates in a plane transverse to the beam direction. p and l are number of radial and angular nodes. L_p^l is the generalized Laguerre polynomial of order p and index l. That is

$$L_0^1(\rho) = 1, \quad L_1^0(\rho) = 1 - \rho \; ; \; L_2^0(\rho) = 1 - 2\rho + (1/2)\,\rho^2 \qquad \ldots\ldots\ldots\ldots[5]$$

Where $\rho = 2\,r^2(z)\,/\,w^2(z)$

Radial intensity distribution and field configuration of various modes in a resonator are illustrated in figure-2 and -3 respectively.

4.4 FUNDAMENTAL MODE – GAUSSIAN BEAM

A light beam in a "Fundamental mode TEM_{oo}" is known as Gaussian beam. The decrease of the electric field amplitude E(r) and its intensity / power I (r) with distance from its propagation axis is given by

$$E\,(r) = E_0\,\{w_0\,/w(z)\}^2\;\exp -\{r/w(z)\}^2 \qquad \ldots\ldots[6]$$

$$I\,(r) = I_0\{w_0\,/w(z)\}^2\;\exp-2\{r/w(z)\}^2 \qquad \ldots\ldots..[7]$$

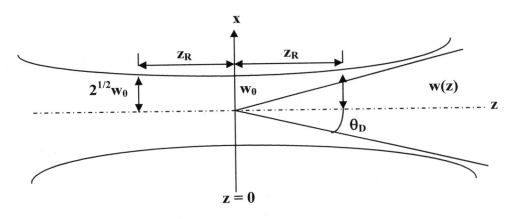

Figure-4 Gaussian Beam Geometry

w- the quantity in radial distance where the electric intensity drops to 1/e and the power density is decreased to $1/e^2$. The parameter w is often called beam radius or "Spot Size". The fraction of total power of a Gaussian beam passing through an aperture with radius r = w, r = 1.5w and r = 2 w is 86.5%, 98.1% and 99.996. Only 10^{-6}% of beam power is lost in aperture of radius 3w.

The Gaussian beam has minimum waist w_0 in the cavity where the wave front is plane as shown in figure-4.

The variation of beam radius w, along the propagation axis, z, can be written as

$$w(z) = w_0 [1 + (z/z_R)^2]^{1/2} \qquad \qquad \dots\dots\dots[8]$$

where $z_R = \pi w_0^2/\lambda$, is known as Rayleigh range or confocal parameter $b = 2 z_R$

The wave front radius of curvature $R(z)$ is given by relation

$$R(z) = z [1 + (z_R/z)^2] \qquad \qquad \dots\dots\dots[9]$$

Wave front has maximum curvature or radius R has minimum value $2z_R$ at $z = z_R$

Half cone angle θ_D and full divergence angle θ of the Gaussian beam (fundamental mode) i.e. $z \gg z_R$ is given by relation

$$\theta = 2\,\theta_D = 2\,[w(z)/z] = 2\lambda/\pi w_0 = 1.27\lambda/(2w_0) \qquad \dots\dots[10]$$

Thus for smaller waist size of beam divergence is more. If R is the radius of curvature and w is the width at a point on z-axis, then the waist radius can be determined from the relation

$$w_0 = w[1 + (\pi w^2/\lambda R)^2]^{-1/2} \qquad \qquad \dots\dots\dots[11]$$

The two regions i.e. $z \ll z_R$ and $z \gg z_R$ are known as near and far-field region respectively.

4.5 RESONATOR CONFIGURATIONS

Beside resonator with plane parallel mirrors i.e. Fabry-Perot resonator, resonators with spherical mirrors are used. Further the resonator are classified under two headings

4.5.1 Stable Resonator

In stable resonator, bundles of light rays are periodically refocused as it travels back and forth between the two mirrors.

4.5.2 Unstable Resonator

In unstable resonator light is dispersed more and more. Some of commonly configurations used for stable and unstable resonators are show in figure-7.

If R_1 and R_2 is radius of curvature of resonator mirrors with spacing length L, then resonator are classified according to following criteria, for stable resonator the resonator length and mirror curvature must satisfy the condition as given below i.e.

$$0 < (1 - L/R_1)(1 - L/R_2) < 1 \qquad \text{or} \qquad 0 < g_1 g_2 < 1 \qquad \dots\dots\dots[12]$$

90

Where $g_1 = 1 - L/R_1$ and $g_2 = 1 - L/R_2$ ....[13]

This criterion is shown in figure-6, where g_1 and g_2 are drawn as the coordinate axes. All cavity configurations are unstable, until and unless they correspond to points located in the shaded area i.e. enclosed by coordinate axes and the hyperbola $g_1g_2 = 1$

The resonator located along the dashed line oriented at 45° with respect to coordinate axis, have symmetric configurations, i.e. resonator mirrors have same radius of curvature. The diagram is divided into positive and negative branches defining quadrants for which g_1g_2 is either positive or negative. The stable and unstable resonator with plane mirrors or spherical mirrors are used depending upon gain of laser material, energy and beam divergence requirement. The merits and demerits of various configurations will be discussed will be discussed in details under active resonators.

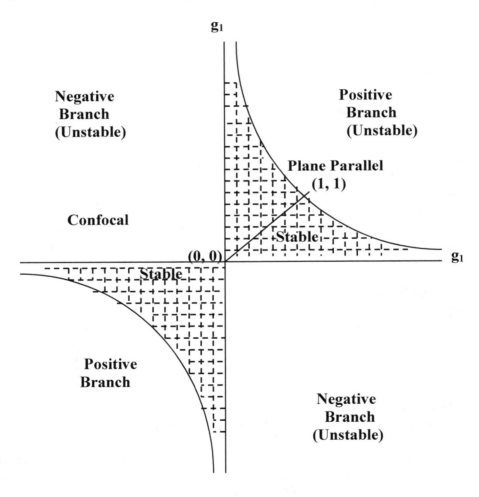

Figure-5 Stable and Unstable Laser Resonator Region

4.6 RESONATORS with MIRRORS HAVING GAUSSIAN BEAM WAVE FRONT CURVATURE

If R_1 and R_2 are the radius of curvature of mirrors corresponding to curvature of Gaussian beam wave front at distance t_1 and t_2 from beam waist w_0 as shown in figure-7, then the radius w_1 and w_2 of beam at mirror R_1 and R_2 as derived by Kogelink and Li [6] are given by

$$w_1^4 = \left(\frac{\lambda R_1}{\pi}\right)^2 \cdot \frac{R_2 - L}{R_1 - L} \cdot \frac{L}{R_1 + R_2 - L} \qquad \ldots\ldots\ldots[14]$$

$$w_2^4 = \left(\frac{\lambda R_2}{\pi}\right)^2 \frac{R_1 - L}{R_2 - L} \cdot \frac{L}{R_1 + R_2 - L} \qquad \ldots\ldots\ldots[15]$$

$$w_0^4 = \left(\frac{\lambda}{\pi}\right)^2 \frac{L(R_1 - L)(R_2 - L)(R_1 + R_2 - L)}{(R_1 + R_2 - 2L)^2} \qquad \ldots\ldots\ldots[16]$$

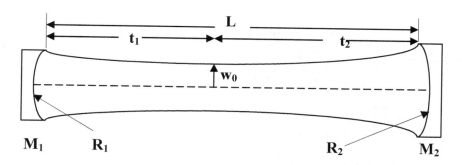

Figure-6 Resonator with mirrors having Gaussian Beam Wave Front Curvature

$$t_1 = \frac{L(R_2 - L)}{R_1 + R_2 - 2L}, \quad t_2 = \frac{L(R_1 - L)}{R_1 + R_2 - 2L} \qquad \ldots\ldots\ldots[17]$$

4.6.1 Resonator with mirrors of equal curvature i.e. $R_1 = R_2 = R$

$$w_{1,2}^2 = \frac{\lambda R}{\pi} \left(\frac{L}{2R - L}\right)^{1/2} \qquad \ldots\ldots[18]$$

$$w_0^2 = \frac{\lambda}{2\pi} [L(2R - L)]^{1/2} \qquad \ldots\ldots\ldots[19]$$

beam waist is at center of resonator i.e. $t_1 = t_2 = L/2$. If $R \gg L$ then

$$w_0^2 = w_{1,2}^2 = \frac{\lambda}{\pi} (RL/2)^{1/2} \qquad \qquad ......[20]$$

4.6.2 Confocal Resonator R = L

$$w_{1,2}^2 = (\lambda R/2)^{1/2} , \quad w_0 = w_{1,2} / 2^{1/2} \qquad \qquad ....[21]$$

i.e. it has smallest mode volume

4.6.3 Plano-Concave Resonator i.e. R_1 = Infinity

$$w_1^2 = w_0^2 = (\lambda / \pi) [L(R_2 - L)]^{1/2} \qquad \qquad ......[22]$$

$$w_2^2 = (\lambda / \pi) R_2 [L / (R_2 - L)]^{1/2} \qquad \qquad ....[23]$$

Beam waist occur at the flat mirror i.e. $t_1 = 0$, $t_2 = L$,

The mode volume of these resonators is illustrated in figure-7, as shaded area. The resonator with R >> L has large mode volume, while spherical resonator has least mode volume.

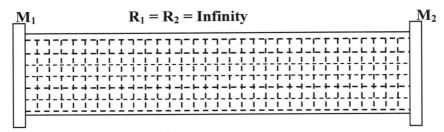

[a] Plane Parallel (Stable)

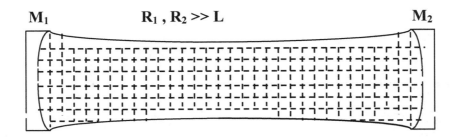

[b] Large Radius Mirror (Stable)

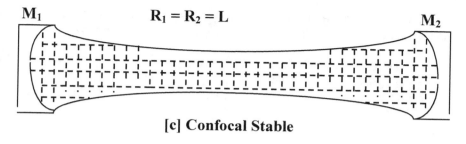

R₁ = R₂ = L appears as $R_1 = R_2 = L$

[c] Confocal Stable

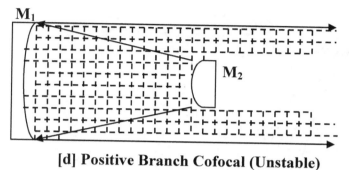

[d] Positive Branch Cofocal (Unstable)

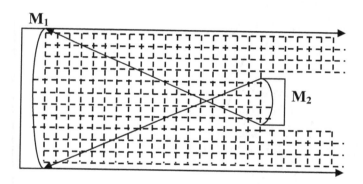

[e] Negative Branch Confocal (Unsable)

Figure-7 Resonator Configuration Showing Mode Volume

4.7 DIFFRACTION LOSSES

The first reasonable approach was given by Li and Fox [2] i.e. numerical calculation using computer. Their approach was to consider a propagating wave reflected back and forth by a pair of plane parallel mirror, is equivalent to the case of transmission media comprising of a series of collinear, identical aperture cut into parallel and equally spaced black partition of infinite extent. A parameter, which was useful in the calculation, is quantity so called Fresnel number.

$$N = a^2 / \lambda L$$

$$\dots\dots\dots\dots[24]$$

Defined for a circular mirror of radius a and L is the distance between the mirrors. The diffraction loss in a resonator increases with decrease in N value.

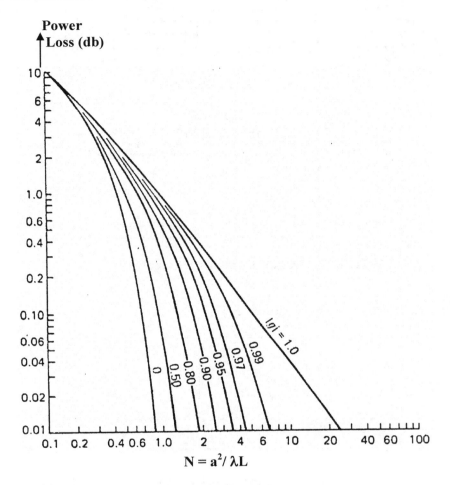

Figure-8: Diffraction Losses per Transit for TEM$_{00}$ Mode for Stable Resonator with Plane and Spherical Mirrors

The general relation for diffraction losses (α) derived in [7] for resonator; with large Fresnel number (N) for plane parallel circular mirrors is given by the following relation,

$$\alpha = 8 \, k_{pl}^{2} \, \frac{\delta(M + \delta)}{[(M + \delta) + \delta^{2}]^{2}} \qquad \qquad \dots\dots\dots\dots[25]$$

Where $\delta = 0.824$, $M = (8\pi N)^{1/2}$, k_{pl} is the $(p + 1)^{th}$ zero of the Bessel function of order l.

The diffraction losses per transit versus Fresnel number N for fundamental mode for stable resonator with plane and curved mirror are plotted in figure-8. The confocal resonator has least losses. The detail analysis of diffraction losses is given in reference [9] for resonator with rectangular and circular mirrors for higher order modes, losses are more due to large beam divergence for higher order modes.

4.8 ACTIVE RESONATORS

In order to compensate reflection and diffraction losses, resonator needs a active medium which may be laser material, solid state laser material excited by intense flash lamp or gas in an electric discharge tube with glass windows or semiconductor junction. Mirrors may be replaced by porro or corner cube prism. Also passive and active Q-switching elements, aligning wedges, polarizing beam splitters or collimating telescopes are introduced in the resonator depending on laser requirement or to compensate distortion introduce by these active elements. Some of the active resonators in stable and unstable configurations used in range finders are as follows.

4.8.1 STABLE RESONATOR CONFIGURATIONS

4.8.1.1 Resonator with Plane Mirror and Porro Prism

This type of resonator is most simplest and very efficient for Nd:Glass, Er:Glass or Ruby lasers. The mirror is plane dielectric coated for 50% to 80% reflectivity, or two or three sapphire plate resonant reflector with peak reflectivity of 66% or 84%. The porro prism act as total reflector whose pyramidal error and right angle accuracy is within 2 seconds. The active material is in the form of clad rod as shown in figure-9a. Antireflection coatings are given to surfaces to reduce losses. Optical materials used are high quality i.e. interferometric grade.

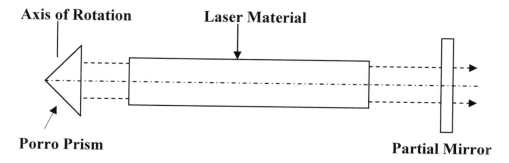

Figure-9a: Resonator with Porro Prism and Plane Mirror

4.8.1.2 Resonator with Plane Mirror and Corner Cube Prism

This type of resonator is shown in figure-9b. This type of resonator is mainly used for Nd:YAG laser material with passive Q-Switching material inside the cavity. The plane mirror has 50% reflecting coating or variable reflecting coating on one end of laser rod. With use of corner cube prism of high angle accuracy small beam deviation due to passive Q-Switched element do not disturb the alignment of resonant mirrors [10].

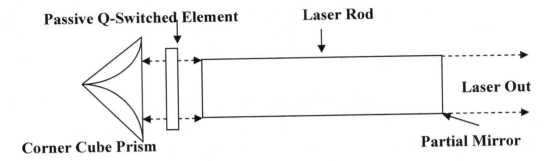

Figure-9b: Resonator with Corner Cube Prism and Mirror

4.8.1.3 Resonator with Porro Prisms and Polarizing Beam Splitter

This type active resonator is shown in figure-9c. The porro prisms act as total reflectors, while the output is taken through the polarizing beam splitter. The electro-optic crystal is also placed in the resonator for Q-Switching. Some time beam-expending telescope is placed inside the resonator to reduce energy density on electo-optic crystal and at the same time it can compensate distortion produce in laser rod when it is pumped by cw-source or at high repetition rate. Aligning wedges are also introduced for fine alignment of mirrors. The laser generally uses Nd:YAG crystal.

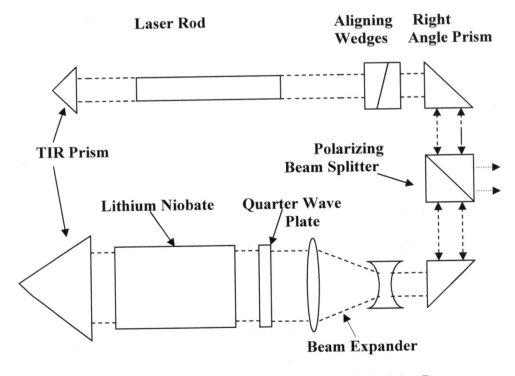

Figure-9c: Resonator with Porro Prism, Beam Expander & Polarizing Beam Splitter.

4.8.1.4 Folded Zig-Zag Slab Resonator

This type of resonator as shown in figure-9d, is used in a diode side pumped laser crystal in the form of a slab. Here, interaction of feedback light flux with laser crystal is four times more as compared to linear resonator. An electro-optic Q-Switching arrangement is placed between laser slab and partial mirror. This type of laser is very efficient and very small heat is dissipated in laser from diode pump source. This type of resonator is shown in figure-9d. This type of resonator can be used for high repetition rate ranging system only using conduction cooling.

Partial Reflector   **Nd:YAG Slab**

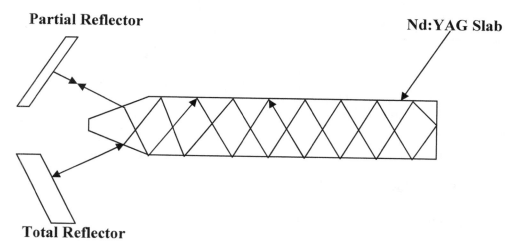

Total Reflector

Figure-9d: Co-Planar Folded Zig-Zag Plane Resonator

4.8.2 UNSTABLE RESONATOR CONFIGURATIONS:

Unstable resonator configurations [3, 14] are used for lasers whose materials have medium and high gain i.e. like Nd:YAG and CO_2 laser systems. In these configurations, the output laser beam in fundamental mode is obtained even with large Fresnel number i.e. with large laser diameter and short resonator length. The two configurations most commonly used for range finders are

4.8.2.1 Confocal Positive Branch Unstable Resonator

This type of resonator is use for very high power/energy laser as shown in figure-10a. The resonator magnification M is the ratio of outer D to inner diameter d of laser beam i.e.

M = D/d

.........[26]

The geometrical output coupling is related to the magnification M by

$$T = 1 - \frac{1}{M^2} \qquad \qquad \dots\dots[27]$$

Ignoring internal resonator losses, the round-trip gain G of the laser has to be

$$G = \text{or} > M^2$$

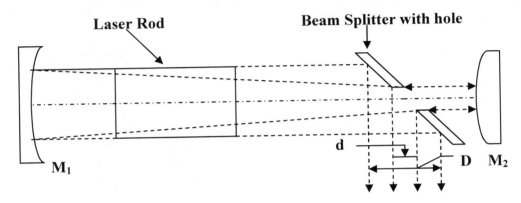

Figure-10a: Positive-Branch Unstable Resonator

For a confocal resonator, the mirror radii are given by

$$R_1 = \frac{-2L}{M-1}, \quad \text{and} \quad R_2 = \frac{2ML}{M-1} \qquad \dots\dots[28]$$

Where L is the resonator length and R_1 and R_2 are the output and back resonator mirror curvatures. The detail design is given in reference [15]

4.8.2.2 Negative- Branch Unstable Resonator

This type of resonator is shown in figure-10b. This type of resonator is preferred in lasers with peak power less than 10 MW due to focal point in the cavity. The advantage with this type of resonator is that its performance is not significantly degraded with a mirror misalignment angle of as much as a few mill radians. The design parameters for this type of resonator are

$$R_1 = 2L / (M + 1) \quad \text{and} \quad R_2 = 2 M L / (M + 1) \qquad \dots\dots[29]$$

Where L is the confocal resonator length and M is the optical magnification defined before. The details are given in references [13, 14]. This type of resonator is very useful for compact laser range finder cum designator for military applications, since the output power is not significantly degraded, even with a mirror misalignment angle of as much as few mill radians [14].

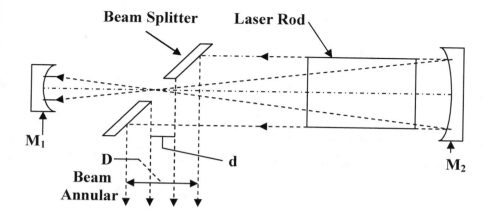

Figure-10b: Negative-Branch Unstable Resonator

4.8.3 RESONATOR WITH RADIALLY VARIABLE REFLECTING MIRRORS

The generation of high power diffraction limited laser beam using positive or negative branch unstable resonators have drawback; firstly, the near field pattern of beam is in annular shape. Secondly, the negative branch resonator has a beam focussing point inside the resonator, limiting its use unto certain power. These drawbacks have been overcome using radially variable reflecting mirrors. The variation in reflectivity is a Gaussian or Super Gaussian function, therefore these mirrors of resonators are known as Gaussian / Super Gaussian mirrors [18,19] as shown in Figure-11.

4.8.3.1 Resonator With Gaussian Mirrors

The Gaussian mirror has the following shape of radial intensity reflection profile

$$R = R_0 \exp [- 2 (r/w_m)^2]$$ ....[30]

Where R_0 is the peak reflection at $r = 0$ i.e. r is the radial co-ordinate and w_m is spot size at mirror as shown in figure-11.

The magnification of resonator with Gaussian mirror is defined as

$$M = w_i / w_r$$ ....[31]

Where w_i and w_r are the spot size of incident and reflected beam at Gaussian mirror.

The effective reflectivity, R_e of the mirror for TEM_{oo} is given by

$$R_e = R_0/M^2$$ ....[32]

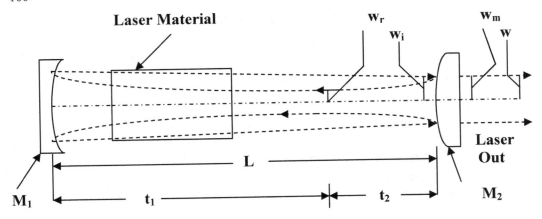

Figure-11 Laser Resonator with Output Mirror Reflectivity Variable in Gaussian / Super-Gaussian Profile.

The spatial profile of the transmitted beam is the product of the incident Gaussian beam and the transmission profile of the mirror and is given by

$$I_{out}(r) = \{1 - R_0 \exp [-2(r/w_m)^2]\} \, I_0 \exp [-2(r/w)^2] \qquad \text{......[33]}$$

The intensity profiles for mirrors of various peak reflectivities are given in [18].

If R_0 is equal to $1/M^2$, output beam profile is flat. It has depression for values greater than $1/M^2$.

The output beam coupling T of resonator is

$$T = 1 - 1/M^2 \qquad \text{............[34]}$$

4.8.3.2 Resonator with Super Gaussian Mirrors

The super Gaussian reflectivity R(r) profile is defined by

$$R(r) = R_0 \exp [- 2(r/w_m)^n] \qquad \text{......[35]}$$

Where R_0 is the peak reflectivity at $r = 0$ and w_m is the mirror spot size or radial distance at which the reflectivity falls to $1/e^2$. The order of reflectivity profile $n > 2$ is for Super Gaussian mirror.

The spot size w_I of the incident beam onto the Super Gaussian mirror is given by

$$w_i = w_m (M^n - 1)^{1/n} \qquad \text{............[36]}$$

The resonator magnification is given by

$$M = w_i / w_r \qquad \qquad \dots\dots\dots[37]$$

Where w_r is the spot size of reflected beam.

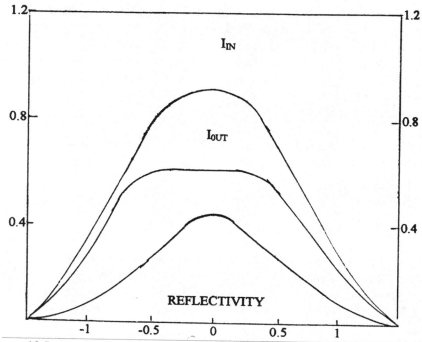

Figure-12 Intensity Profile of Incident and Transmitted beam with Output Mirror Reflectivity Variable in Super-Gaussian Profile (n = 2.8, M = 1.6, R_0 = 0.44).

The profile of transmitted beam is given by

$$I_{out}(r) = \{\ 1 - R_0 \exp\ [-2\ (r/w_m)^n]\}\ I_0 \exp\ [-2(r/w)^n] \qquad \dots\dots[38]$$

The maximum flat output beam is obtained when

$$R_0\ M^n = 1 \qquad \qquad \dots\dots\dots[39]$$

If $R_0\ M^n > 1$, then center portion of beam has depression as shown in figure-12 i.e. for n=2.8, R_0 = 0.44 for uniform output beam profile.

4.9 LONGITUDINAL MODES – SINGLE FREQUENCY OPERATION of LASERS

Lasers operating in a fundamental mode TE_{00p} in a resonator have number of axial modes, where p is an integer and can takes values such that

$$L = p\ (\lambda/2) \qquad \qquad \dots\dots\dots[40]$$

Where L is resonator length.

If active material is introduced in the cavity, the value of p is limited to wavelength or frequency that depends on gain and fluorescence line width of laser material. Fluorescence line width is very sharp for gas laser as compared to solid-state lasers. Therefore, numbers of longitudinal modes in the gas laser output are smaller in number as compared to solid-state lasers. In case of semiconductor gain bandwidth is large, but due to small cavity length, number of axial modes are limited. As the coherence length depends on these modes, for single frequency operating laser, use of Etalon in resonator or resonant reflector of high refractive index optical material as partial mirror can restrict operation of these modes.

For fundamental mode beam, separation of wavelength $\Delta\lambda$ or frequency $\Delta\nu$ between two adjacent modes is given by

$$\Delta\lambda = \lambda_0^2/2L \qquad \text{or} \qquad \Delta\nu = c/2L \qquad\qquad\qquad\ldots\ldots\ldots\ldots[41]$$

4.9.1 Fabry Perot Etalon

A Fabry Perot Etalon is a plate of thickness d whose surfaces are parallel to each other. When a ray of light as shown in figure-13, enters the plate at incident angle θ, multiple reflection occurs. If two reflecting components at M_1 destructively interfere, then maximum energy is transmitted. If they are in phase, we have maximum reflection [20].

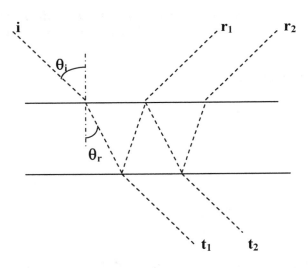

Figure-13 Fabry-Perot Etalon

The phase difference between these components is given by

$$\delta = 2\pi/\lambda \,(2\mu d \cos\theta) \qquad\qquad\qquad\ldots\ldots\ldots\ldots[42]$$

Where μ is refractive index.

Maximum value of transmission takes place when

$$2\mu d \cos\theta = m\lambda, \quad \mu = 1, 2, 3, \ldots\ldots \tag{43}$$

Minimum transmission/ maximum reflection takes place at

$$2\mu d \cos\theta = m\lambda/2, \quad m = 1, 3, 5 \ldots \tag{44}$$

The maximum value of reflection is

$$R_{max} = 4\, r\,/\,(1+r)^2 \tag{45}$$

Where r is reflectivity of each surface given by

$$R = (\mu - 1)^2/(\mu + 1)^2 \tag{46}$$

4.9.2 Resonant Reflector

Relation [46] gives peak reflectivity of an uncoated etalon with high optical refractive index material used as partial mirror.

The maximum reflectivity of a multi-plate resonant reflector is

$$R_{max} = (\mu^{N'} - 1)^2/(\mu^{N'} + 1)^2 \tag{47}$$

Where N' is the number of reflecting surfaces.

For resonant reflector with two sapphire plates i.e. $\mu = 1.79$, $N' = 4$,

$$R_{max} = 0.67 \tag{48}$$

The resonant reflectors are used with Q- Switched solid-state lasers, operating in very high gain mode.

The peak reflectivity for single, two and three plate sapphire reflectors are 0.25, 0.66 and 0.87 respectively.

The design of multiple plate resonators can be optimized with the aid of computer program [17] for plate width and plate spacing should be such that desired wavelength peak reflectivity to center of fluorescence line where gain of laser material is maximum.

4.10 SINGLE AXIAL MODE OPERATION-LINE WIDTH

According to Schalow-Townes [21], finite line width of laser operating in a single axial mode is

$$\Delta v_L = \frac{2\pi h v (\Delta v_c)^2}{P_{OUT}} \qquad \ldots\ldots\ldots[49]$$

Where Δv_c is the line width of the passive resonator, hv is the photon energy and P_{OUT} is the laser output power.

If τ_c is lifetime of the photon in the resonator, then

$$\Delta v_c = 1 / 2 \pi \tau_c \qquad \ldots\ldots\ldots[50]$$

Therefore from these relations, we have

$$\Delta v_c = h v / 2\pi\tau_c^2 P_{OUT} \qquad \ldots\ldots\ldots[51]$$

If n is the number of photon in a resonator, then relation gives laser output power

$$P_{OUT} = n h v / \tau_c \qquad \ldots\ldots\ldots[52]$$

Therefore from these above relations we have

$$\Delta v_L = \Delta v_c / n \qquad \ldots\ldots\ldots[53]$$

Therefore from this relation laser line-width can be a fraction of milli-hertz. In actual practice line width is in few kilo-hertz to hundreds of mega-hertz and fluctuations in frequency due to change in cavity length and refractive index caused by acoustical and mechanical vibrations, power supply instabilities, variation of temperature and pressure etc. Single mode operation of diode pumped Nd:YAG laser i.e. mini lasers [22]. In semiconductor laser, single mode operation is achieved by coupled cavities or frequency selective feedback [21]. However, with active methods and precise temperature control, operation of laser has been achieved in milli-hertz bandwidth.

SUMMARY

1. Optical resonator theory is basis of laser development. Main source of light loss in a resonator is diffraction.

2. Laser output from resonator is in form of discrete frequencies and a pattern known as longitudinal and transverse modes.

3. A light beam in a fundamental mode is known as Gaussian beams.

4. Spherical resonators have less diffraction losses as compared to Fabry-Perot resonator, but their mode volume is less and in the cavity, whole excited laser material does not contribute for laser output.

5. For high gain laser materials unstable resonators have advantages of large laser mode volume and better beam quality.

6. A mirror with variable radial reflectivity gives more uniform output in a fundamental mode. These mirrors are known as Gaussian/ Super Gaussian mirrors.

7. Single axial mode operation is possible if cavity length is reduced and at the same time etalon or resonant reflector or cleaved coupled cavities are used.

8. With active methods, frequencies with bandwidth of milli hertz have been achieved in a single axial mode as indicated in Towns and Schalow relation.

LIST OF SYMBOL USED

a   = Radius of Resonant Mirror.

c   = Velocity of Light.

D, d   = Outer and inner diameter of annular laser beam.

$E(x, y)$ = Electric field distribution in x and y co-ordinates.

h   = Universal Plank's Constant.

$I_{nm}(x,y,z)$ = Light intensity at x, y and z co-ordinates.

$I_{pl}(p,l,z)$ = Light intensity at r, ϕ, and z in polar co-ordinates.

$H_n(x)$ = Hermite polynomial of nth order.

K_{pl} = $(p + 1)^{th}$ Zero of the Bessel function of order l

L   = Resonator length.

L_p^I = Generalized Laguerre polynomial of order p and index I.

M   = Resonator magnification.

N   = Fresnel number.

N'   = Number of surfaces of resonant reflector.

n   = Number of photon in a resonator.

P_{out} = Laser output power.

R_1, R_2 = Radius of curvature of cavity mirrors.

R_0 = peak reflection of variable radial reflecting mirror at r = 0.

$R(z)$ = Radius of curvature of Gaussian beam at co- ordinate z.

R_{max} = Maximum peak reflectivity of resonant reflector.

R_w = Effective reflectivity of variable reflectivity mirror in Gaussian profile.

R   = Reflectivity of output mirror.

T   = Transmission of output mirror.

t_1, t_2 = Distances of cavity mirrors from Gaussian beam waist.

TEM_{mnp} = Transverse electric mode with n and m nodes in x and y direction with p number of half-Wavelengths in cavity of length L.

TEM_{plq} = Number of electric nodes in r, θ and z direction.

w_0 = Minimum spot size/beam waist of Gaussian beam.

w_1, w_2 = Spot size of Gaussian at cavity mirrors.

$w(z)$ = Spot size of transverse beam in which intensity decrease to $1/e^2$ of the peak intensity.

w_i, w_r = Spot size of incident and reflected at cavity mirror whose reflectivity varies in Gaussian or super Gaussian way.

w_m = Radial distance at which reflectivity decreases to $1/e^2$ of its maximum value at center.

ν = Frequency of light.

$\Delta\nu$ = Frequency spacing between adjacent longitudinal modes $\Delta\nu_L$ = Line width of laser operating in single axial mode.

$\Delta\nu_c$ = Line width of the passive resonator.

τ_c = Life time of the photon in the resonator.

μ = Optical refractive index of etalon

REFERENCES

1. Schawlow, A. L., and Townes, C. H., " Infrared and Optical Masers", Phys. Rev., Vol. 112, December 1958, pp. 1940-1949.

2. Fox, A. G., and Li, T., "Resonant Modes in a Maser Interferometer, " Bell Sys. Tech. J., Vol. 40, March 1961, pp. 453-458.

3. Siegma, A. E., "An Introduction to Lasers and Masers", Mc Graw Hill Book company, New York (1971).

4. Gould, G., Phys. Rev, Vol. 101, 1956, pp. 1828.

5. Boyd, G. D. and Kogelink, K., "Generalized Confocal Resonant Theory", Bell Syst. Tech. J., Vol.41, 1962, pp. 1347-1369.

6. Seigma,A.E., "Unstable Optical Resonator for Laser Applications", Proc. IEEE, Vol. 53, No. 3, March 1965, pp. 277-287.

7. Kogelink, K. and Li, T., "Laser Beams and Resonators", Applied Optics (USA), Vol. 5, No.10, October 1966, pp. 1550.

8. Vainshtein, L. A., "Open-Resonator for Lasers", Sov. Phys.- JETP, Vol. 17, September 1963, pp. 709.

9. Vaishtein, L. A., "Open-Resonators with Spherical Mirrors", Sov. Phys.-JETP, Vol. 18, February 1964, pp.471.

10. Edson, R. Peck, "Polarization Properties of Corner Reflectors and Cavities", J. Opt. Soc. Of America, Vol. 52, No. 3, March 1962, pp. 253-257.

11. Gould, G., Jacobs, S., Rabinowitz, P. and Schultz, T., "Crossed Roof Prism Interferometer", Applied Optics (USA), Vol.1, No.4, July 1962, pp. 533-536.

12. Mansharamani, N., "High power Nd:YAG Laser- Light Flux Recirculation of Energy from Residual Inverted Population", presented at National Symposium on High Power Lasers, held on December 23-24, 1996 at Defence Science Centre, Delhi-110054.

13. Richards, J. and McInnes, A., "Versatile Efficient, Diode-Pumped Miniature Slab Laser", Optics Letters, Vol. 20, No. 4, 1995, pp. 371-373.

14. Sakies, P. H., "A Stable YAG Resonator Yielding a Beam of Very Low Divergence and High Output Energy", Opt. Commun., Vol. 31, No. 2, November 1979, pp.189-192.

15. Siegman, A. E., "Unstable Optical Resonators", Applied Optics (USA), Vol. 13, No. 2, February 1974, pp. 353-366.

16. Hanna, D. C., Sawyers, C. G. and Yuratich, M. A., "Large Volume TEM$_{00}$ Mode Operation of Nd. YAG. Lasers". Optical Communication, Vol. 37, No. 5, 1 June 1981, pp. 359-362.

17. Watts, J. K., "Theory of Multi plate resonant Reflectors", Applied Optics (USA), Vol. 7, No. 8, August 1968, pp. 1621-1623.

18. Magni,V., De Silvestri, S. and Cybo-Ottone, C., "Resonator with Variable Reflectivity Mirrors", pp. 94-104, The Physics and Technology of Laser Resonators, Edited by Hall, D. R. and Jackson, P. E., Publisher Adam Hilger, Bristol and New York (1989).

19. Sandro De Silsestri, Vittorio Magni, Orozio Svelto and Valentini, G., "Laser with Super Gaussian Mirrors", IEEE J. Quantum Electronics, Vol. 26, No. 9, September 1990, pp. 1500-1509.

20. Born, m. and Wolf, E. "Principles of Optics", Publisher Pergamn, New York 1964.

21. Abramski, k. M. and Hall, D. R., " Frequency Stabilization of Lasers", pp.117-131, the Physics and Technology of Laser Resonators, Edited by hall, D. R. and Jackson, P. E., publisher Adam Hilger, Bristol and New york, (1989).

22. Das, P. "Lasers and Optical engineering", Publisher Springer-Verlag, Berlin and New York, 1991.

CHAPTER – 5

LASER PUMPING

5.1 INTRODUCTION

The quantum efficiency of a laser mainly depend on pump source i.e. excitation method for population inversion. The laser sources used in range finders as described in chapter-3 are solid state, gaseous and semiconductor. Flash lamps or semiconductor laser diode pumps the solid-state laser sources. The gaseous lasers are pumped by high voltage direct current or radio frequency discharge through gaseous medium. The semiconductor laser operates by forward high current pulses or sinusoidal modulated current to semiconductor laser.

Since from last forty five years, the quantum efficiency of solid state lasers [1] have improved from 1% to 20% by improvement in flash lamps emission to match absorption band of laser material or by semiconductor laser diode output pumping directly into absorption band [2] or using sensitized materials in reflectors or pumping lamp envelope to covert ultraviolet radiation into absorption band. In case of the gaseous lasers from 0.1% in case of He-Ne laser to 60% for CO_2 laser by using resonance pumping using gas mixture. In case of semiconductor lasers from 15% to 95% by improving better interaction between photon and electron-holes in active region i.e. confining electron-holes in very narrow region known as quantum well region surrounded by photon in a graded index hetero-structure.

5.2 PUMP SOURCES FOR SOLID STATE LASERS

As discussed in the chapter-3, solid state laser sources used for ranging application are ruby, Nd:Glass, Nd:YAG or Er:Glass. The major absorption bands for ruby are at 0.404 µm and 0.554 µm, for Nd:Glass are 0.58 µm, 0.75 µm and 0.81 µm, for Nd:YAG are 0.75 µm and 0.81 µm and for Er:Glass are in range of 0.94 to 0.99 µm. Therefore considering the above absorption band and geometry of solid-state laser materials, essential requirements of pump sources are

*Good optical and electrical efficiency in visible and near infrared as indicated in above paragraph.

*Free from ultra violet (uv) radiation, as uv-radiation can cause permanent damage to laser material.

*Rugged with long operating and storage life.

*Geometry of pump source should be such that its output can be easily and uniformly coupled to laser material.

*It should be cheap and can be easily replaceable in system.

*It can be operated efficiently in pulse duration depending on mean lifetime of laser material.

Considering the above requirement, the pump source used for laser pumping are Xenon / Krypton flash lamps [3,4,5,6] or semiconductor quantum well diode lasers depending upon role for which laser range finder is used.

5.2.1 XENON / KRYPTON DISCHARGE LAMPS

Xenon lamps are used for pumping solid-state laser. Numbers of geometry like helical, annular, coaxial and linear shapes have been tried for pumping. But linear flash lamps offer best geometry for coupling output to laser material, which is in the form of rod or slab. The reflectors for coupling light output of linear flash lamp are elliptical, circular in close-coupled configuration. When close-coupled reflectors are used, they are diffuse or silver-plated on outer glass surface of tube. Krypton lamp is used for continuos pumping of Nd:YAG crystal or in pulse discharge for low energy operation as output line spectrum of krypton lamp matches with absorption bands of Nd:YAG.

5.2.1.1 Construction

The linear xenon flash lamp as shown in figure-1a and 1b consist of pure fused silica tubing or fused silica doped with cerium or samarium. This doped material absorbs ultraviolet (uv) and fluoresces in yellow region to increase pumping efficiency and at the same time avoid solarizations of active ions in a laser material. Absorption of 1-micron flux from flash lamp light output reduces heating of laser material and reduction of inverted population does not take place during storing of atoms in excited state for generation of giant pulse in Q-switch operation. The electrodes generally used are of tungsten anode and thoriated tungsten cathode with a low work function, which makes it easier to trigger the lamp. The tungsten rod seal using an intermediate highly doped

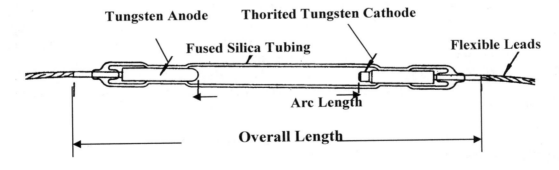

Figure 1(a)-Xenon / Krypton Flash lamp Tungsten Rod Seal.

Anode   **Cathode**

Figure 1(b) - Xenon / Krypton Flash Lamp with Indium Solder End-Cap Seal

borosilicate glass to seal the electrode to the fused quartz envelope. With this type of seal the flash lamp can be heated to 1000^0 C for degassing purpose, which result in for long shelf life. The second type of electrode is brazed to copper rod welded to nickel cup with a low temperature indium solder seal between the copper plated nickel cup and the platinum coated end of the quartz envelope. This type of seal must be operated below 180^0C. This type of seal permits large cross section of electrode permitting coolant to cool tip of electrode for high power operation of flash lamp. As degassing in this type of lamp cannot be done at high temperature, its shelf life is small. These types of lamps are very rarely used, that too only in very high power operation. The gas generally filled is at 450 to 700 torr. Though high pressure filled gasses have better efficiency, but their operating life is less and moreover difficult to trigger.

5.2.1.2 Optical Characteristics

The light emission depends on gas filled its pressure and discharge current density. The output spectrum of flash lamp constitutes of line and continuum spectrum as shown in figure-2a and -2b. The line spectrum of krypton flash lamp matches with absorption spectrum of Nd:YAG laser material. Therefore at low operating power krypton lamps are more efficient than xenon flash lamp [5]. Krypton lamp filled at pressure 700 torr in a size of 3 mm bore and arc length 50 mm operates in pulse energy less than 8 Joules is more efficient than xenon lamp of same specifications. Xenon flash lamp has 40 to 60% efficiency for light between 0.2 to 1µm, while krypton lamp

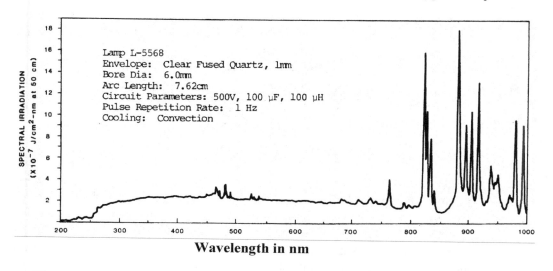

Wavelength in nm

Figure-2(a) Spectral Output of Low Pressure Xenon Flash Lamp (Pressure- 390 Torr) [Courtesy of ILC Reference-5]

112

efficiency is 25 to 30%. Krypton lamps are used for continuous pumping of Nd:YAG, since its emission at 0.73 to 0.76 μm and 0.79 to 0.82μm matches well with main pumping bands of Nd:YAG. As the pressure of gas in the flash lamp is increased, its efficiency is increased.

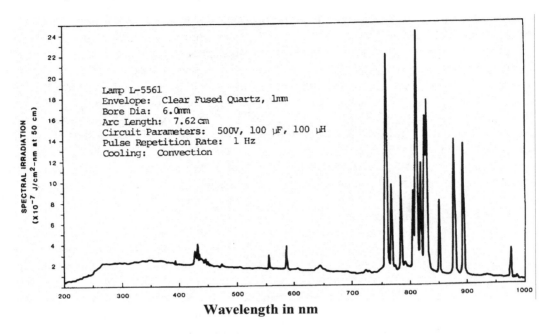

Figure-2(b) Spectral Output of Low Pressure Krypton Lamp (700Torr) [Courtesy of ILC Reference-5].

The gas pressure upto 3000 Torr is used in lamps for certain high power application, but high-pressure lamps are difficult to trigger and its operating life is also less. In pulse operation at high current density beyond 4000 A/cm^2, lamps become opaque for longer wavelength, its emission from center core of lamp is less for longer wavelengths. Thus output of lamps is reduced for infrared, especially lamps with large bore diameter. Xenon flash lamp at a current density of 2500 A/cm^2 has maximum efficiency [7].

5.2.1.3 Electrical Characteristics and Operation of Flash Lamp:

The life and efficiency of lamp depends on electrical operation i.e. type energy storage capacitor, operating voltage, pumping pulse duration, current density, trigger method, series inductance used and lamp impedance [7,8]. The lamp impedance is a variable parameter and is function of current density. Depending upon trigger method, the arc grows from a period varying from 5 to 50 μ second to fill the tube. After the arc is stabilized, the voltage and current relationship describe by relation

$$V = K_0 \, i^{1/2} \qquad\qquad\qquad \dots\dots\dots [1]$$

Where $K_0 = k\,l/d$ and $k = 1.27$ for gas pressure of $P = 450$ torr   ....[2]

For other pressures

$k = 1.27\ (P/450)^{0.2}$ ....[3]

The lamp resistance for gas pressure of $P = 450$ torr   during high current is

$R_L(i) = 1.27\ (l/d)\ i^{-1/2}$ ....[4]

The equivalent circuit during electric discharge through is shown in figure-3, where C is the capacity of energy storage condenser and L is series inductance, R is lamp impedance at time.

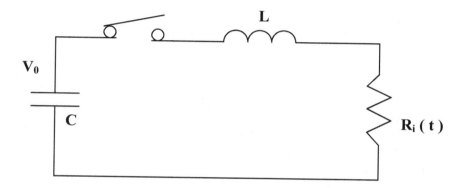

FIGURE 3- Discharge Circuit of Condenser Through Inductance & Flash Lamp.

The discharge of condenser is represented by equation

$$L\frac{d^2q}{dt^2} + R\frac{dq}{dt} + \frac{q}{C} = 0$$ ....[5]

As a solution of this equation the voltage across the energy storage condenser v(t) is given by

$$v(t)\ =\ V_0\ (\alpha/\beta)\ \{\exp.[-(\alpha-\beta)t] - \exp.(\alpha+\beta)t\}$$ ....[6]

$$\text{Where}\ \ \alpha = \frac{R}{2L}\ ,\ \ \beta = \left(\frac{R^2}{4L^2} - \frac{1}{LC}\right)^{1/2}$$ ....[7]

Initially at $t = t_0$, Charge on condenser is

$q_0 = C\,V_0$ and current $dq/dt = 0$ ....[8]

114

Initially at $t = t_0$, Charge on condenser is

For critically damped case $\beta = 0$ $R^2 = 4L / C$ ....[9]

For over damped case $R^2 > 4L / C$ ....[10]

For under damped case $R^2 < 4L / C$ ....[11]

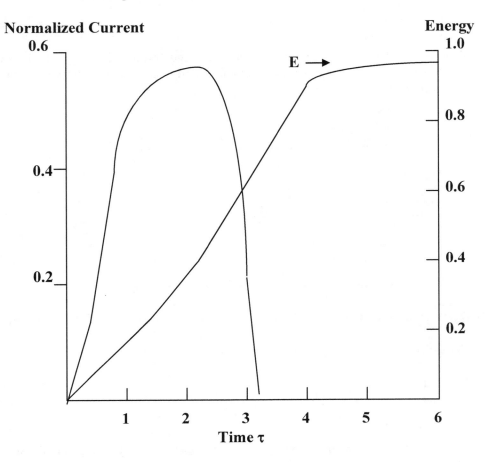

Figure-4 Normalized Current and Energy of a Critically Damped Discharge Circuit.

Considering non-linear load i.e. impedance parameter of lamp K_0, a damping factor α is defined which determine the shape of current pulse. It is

$$\alpha = K_0 (V_0 Z_0)^{-1/2}$$ ....[12]

Where $Z_0 = (L / C)^{1/2}$ ....[13]

Solution of the non-linear differential equation of the lamp current for various values of α have been obtained using computer [8].

As seen from the curve in figure-4, for critically damped case, $\alpha = 0.8$

From the curve, current pulse width at 10% point is

$$t_p = 3T, \text{ where } T = (LC)^{1/2} \qquad \ldots\ldots..[14]$$

Relation gives the electrical energy in Joules on energy storage condenser with capacity C in farad and voltage v_0 in volts

$$E_0 = (1/2) C V_0^2 \qquad \ldots\ldots.[15]$$

Inductance L required for 10% point current pulse i.e. duration in which 97% of energy of condenser is released in flash lamp for damped mode is

$$L = t_p^2 / 9C \qquad \ldots\ldots\ldots.[16]$$

For shorter current pulse width with small series inductance i.e. condition of under damped oscillation, a diode in reverse direction across condenser does not allow condenser to be charged in reverse direction which otherwise reduces operating life of flash lamp and condenser.

Ionizations in a flash lamp is initially produced by shunt, series or simmers trigger method. The operating life and efficiency of flash lamp is more in later two methods, shunt trigger method is used for compact system with very low frequency operation of system with low output energy requirements.

5.2.1.4 Maximum operating / Explosion Energy of Flash Lamp

The maximum operating of flash lamp is 40% of the explosion energy. Explosion energy is energy, in which shock wave and plasma heat causes rupture of the lamp walls and consequently destroy the lamp. The maximum operating depends on current pulse width and cooling method used i.e. flash lamp in free air convection cooling can handle $5W/cm^2$, while for same flash lamp, liquid cooling permits operation of lamp with surface loading of 300 W/cm^2. Almost 50% of electrical energy supplied to lamp is dissipated as heat at the tip of electrodes, 25% as shock wave and in heating the walls of lamp and rest 25% as radiant energy which ends in pumping laser material and heating of reflector and laser material. For high power operating flash lamp electrode shape is such that heat is transferred to walls of flash lamps from tip of electrodes before heat reaches to glass seal point, since walls of lamp can be easily cooled by forced air or liquid coolant.

Relation gives the explosion energy of lamp

$$E_{exp.} = K_2 . l . d . t_p^{1/2} \qquad \ldots\ldots..[17]$$

Where K_2 is function of gas and fill pressure
l = arc length

d = bore diameter of glass envelope

$t_p` $ = pulse width for 1/3 of peak light output

5.2.1.5 Flash Lamp Failure

The flash lamp failure and hence its life depends on the following factors

*Less trigger energy causes intense shock wave, which creates fine cracks in wall of lamp.

*Very high voltage shunt trigger pulse causes puncher in tube wall near electrodes.

*High discharge current in very short pulse causes melting of quartz envelope from inside of wall, thus walls become opaque due to recrystallisation.

*Insufficient heat dissipation from electrodes causes sputtering of electrodes, thus reducing light output due to opacity of lamp wall.

* Improper mounting causes breakage of glass wall near electrode seal.

*Unclean glass envelope or cleaning of glass wall with solvent which leaves residue causes embedding of dirt or residue in walls of lamp which result in decrease in light output.

Thus cleaning of lamp with proper recommend solvent, stain free mounting of lamp, proper trigger requirement, electric discharge energy in pulse of proper duration as recommended by lamp manufacture, cooling of lamp properly according to power input to lamp will enhance the operating life of flash lamps.

5.2.1.6 Lamp Reflectors

The shape of the reflector depends on flash lamp and laser material shape. Materials for reflecting coating on reflector depends on absorption band of laser material. Generally laser material used are in bar / rod geometry, therefore reflectors used are cylindrical with single or double elliptical cross-section or circular for close coupling as shown in figures-5a and -5b.

(a) The Elliptical Reflectors

Three types of reflectors are generally used in source for range finding or in an application where a medium power laser source in continuous mode or in high pulse repetition rate is required. Double ellipses with two flash lamps on two foci of ellipse on with laser rod on common foci are used, where pumping power requirement is high. This type of configuration illuminates laser rod uniformly, though efficiency is less than single ellipse. Single ellipse with lamp and laser rod on each foci are used for high gain laser material like Nd:YAG. The reflector material is of brass with gold plating when water glycol mixture is used for cooling laser rod and lamp, as gold is inert, do not tarnish easily and has good reflectivity in pumping band of Nd:YAG. Half-single ellipse as shown in figure-5c with laser rod and flash lamp mounted on gold plated brass plate. This

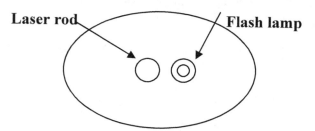

2a = 50 mm, 2b = 48mm, 2c = 14mm

Figure -5a Single Elliptical Reflector

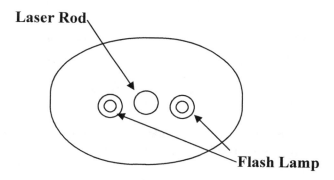

2a =50mm, 2b = 48mm, 2c = 14mm

Figure 5b-Double Elliptical Reflector

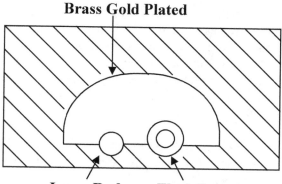

2a = 50mm, 2b = 48mm, 2c = 14mm

Figure 5c - Half – Elliptical Reflector

type of system is used for conduction cooling of lamp and laser rod. For low repetition rate laser chemically coated glass reflector with protection layer gives better efficiency as silver is more reflecting in visible than gold. It has been reported [13] that 90% more efficiency has been achieved by coating aluminum oxide-silver-aluminum oxide layers on Pyrex glass tubing.

The light coupling efficiency from lamp with radius r_L and laser rod radius r_R is given by relation

$$\eta_e = \pi^{-1} [\alpha_0 + (r_R / r_L) \theta_0] \qquad \ldots\ldots\ldots[18]$$

$$\text{Where } \cos \alpha_0 = e^{-1} [1 - 0.5 (1 - e^2) (1 + r_R / r_L)] \qquad \ldots\ldots\ldots[19]$$

$$\sin \theta_0 = (r_L / r_R) \sin \alpha_0 \qquad \ldots\ldots\ldots[20]$$

$$e = c/a \qquad \ldots\ldots\ldots[21]$$

$2c$ = focal separation, $2a$ = major axis and $2b$ = minor axis

$$c = [a^2 - b^2]^{1/2} \qquad \ldots\ldots\ldots[22]$$

(b) Diffuse Reflector

The elliptical reflectors with gold plating or glass reflector with silver coatings are being replaced by barium sulfate diffuse reflector as shown in figure-6 due to following advantages

- Can be used in very close coupling – gives uniform light flux to laser rod, clad or polished cylindrical surface can be used. Life of laser material increases when pumped with uniform light flux

- The glass tubing material of diffuse reflector is doped with ultraviolet (uv) absorbing elements like cerium and samarium – these materials convert uv light flux into useful pump band flux. Thus lamp with pure fused quartz material can be used, these lamps have more operating life and can handle more power as compared to cerium or titanium doped fused quartz. Moreover life of laser material is enhanced as pump light flux is free from uv radiation, with uv radiation   solarization takes place in laser material.

- With clad laser rod having small core diameter of laser material gives uniform laser output in fundamental mode. Thus better ranging capability is achieved with better-collimated beam.

- The life of diffuse reflector is much more than the reflector with metallic coating that gets tarnish with time.

Construction

There are various shapes of diffuse reflector, but most commonly used is shown in figure-6. There is separate bore for flash lamp and laser rod. Advantage with this type is that uv flux from direct radiation not only filtered, but it is converted into useful pumping band also. The outer portion is filled with pure barium sulfate powder with plastic sleeving. Liquid coolant flows throw supporting end plates, over surface of laser rod and flash lamp [14].

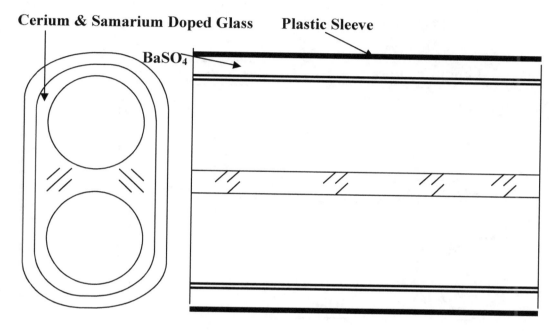

Cerium & Samarium Doped Glass   Plastic Sleeve

BaSO₄

Figure 6 - Diffuse Reflector

5.2.1.7 Cooling Systems

Depending upon the energy input to the flash lamp, three type of cooling systems are used i.e. convection type, conduction type with forced air cooling and liquid cooling system using pure deionised water glycol mixture. The convection system is used, where the input energy to flash lamp is 25 joules in pulse duration of 100 μ second or more with pulse repetition rate of maximum of 12 pulses per minute maintaining a duty cycle of three minutes with seven minutes rest.

Conduction cooling with forced nitrogen in a high-pressure nitrogen filled chamber i.e. five atmosphere or more is used where the pulse input energy is six joules to krypton lamp with arc length of 50 mm in a pulse duration of 100 μ second typically for Nd:YAG system as shown in figure-5c with a pulse repetition rate of ten pulses per second. Here half-elliptical gold plated with flash lamp and laser rod mounted on a thick gold plated brass plate.

120

The liquid cooling system is shown in figure-7. The system consists of magnetically coupled liquid circulating pump, reservoir, and heat exchanger with fan, and deionizer cartridge. The system can be used with any type of power depending upon its size and velocity of liquid over laser rod and flash lamp. But the compact system of range finger can cool laser rod and flash lamp operating at pulse rate of 20 pulses per second with input energy of 10 J in a pulse duration of 100 μ second. This system can be used with gold plated brass elliptical chamber, but diffuse chambers are preferred these days because not only they are more efficient, but also uv content is negligible which does not load deionizer cartage. Photo-1 shows closed liquid cooling system using water glycol mixture developed at IRDE.

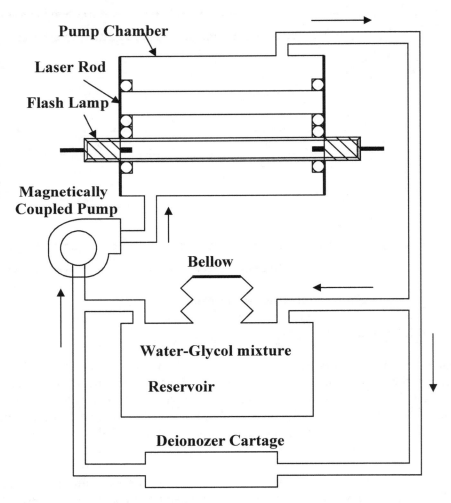

Figure 7- Closed Liquid Cooling System (Water-Glycol Mixture)

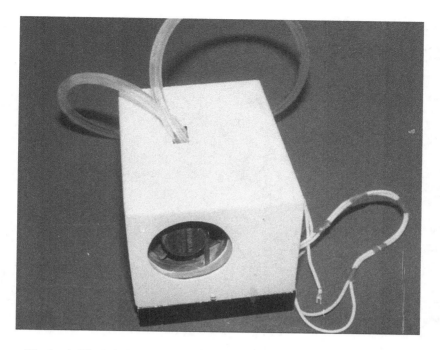

Photo-1 Closed Water & Glycol Mixture Cooling Unit (IRDE)

5.2.2 SEMICOUNDUCTOR QUANTUM WELL DIODE PUMP LASERS

Due to confinement of carrier in a narrow stripe region surrounded by a double heterostructure where radiation are confined as discussed in chapter 3, the efficiency of lasers have increased so that these lasers can be operated in pulse and cw mode. Further reduction in thickness of active region to a thickness of the order of de-Broglie wavelength, the electrons are unable to escape from this region known as quantum well region. This has resulted in drastic reduction in threshold current with the result overall efficiency of these lasers has increased to 50%. Since the output of these lasers can match with the pump band of solid state lasers, the overall efficiency of solid state lasers have increased to 25% instead of 5% when pumped by flash lamp.

The techniques for growing thin layers are vapor-phase epitaxy, the material to be deposited is transported as part of a gaseous compound, a halogen such as gallium chloride ($GaCl_3$), or a tri methyl gallium [$(CH_3)_3Ga$]. When the vapor touches the substrate, it reacts, depositing the material to grow the crystal. Finally, in molecular-beam epitaxy, the substrate is placed in a high vacuum and bombarded with a beam consisting of the material to be deposited. Very thin layers can be grown in this manner. Finally uv or x-ray lithography is used. Molecular beam epitaxy, for AlGaAs / GaAs crystal growth, permits accurate deposition of very thin layer of material.

Generally stripe shape layers are used for more output. Large stripe layers are divided into a row of narrow stripe by blocking with insulted layer for confinement of

122

current for uniform gain i.e. gain- guiding. Further index guiding of radiation in a narrow region results in a better efficiency of lasers as shown in figure-8.

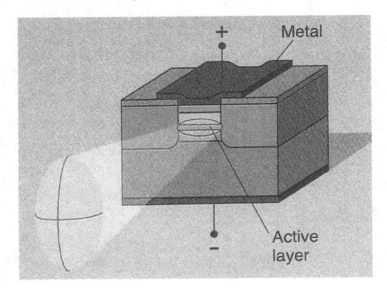

Figure-8: Double Hetero-structure Index Guided Stripe Laser

M/S SDL Inc, of USA is manufacturing and marketing high brightness GaAlAs laser diode partially coherent with output of 4.0 W cw emission area of 1X500 μm. The output wavelength is at 808 nm with 2 nm spectral widths. Electrical to optical conversation efficiency is 30% typical. The output at this wavelength is used to pump Nd:YAG laser material. The end pumping is preferred as emission is from small area as shown in figure-9. A 20 Watts, 10mm array of photo diode is used for side pumping Nd:YAG slab as shown in figure-10. This SDL-3400 array can be cooled with conduction cooling. But liquid cooling is preferred for better life of system. The array is provided with monitoring photodiode in order to achieve constant output power and also to keep array under safe operating conditions. The safe operating temperature of diode array is between –20 to 30^{0}C. Cr^{4+}:YAG is generally used as passive Q-switched element for pulse operation of these lasers for better operating life.

InGaAs diode lasers operating at wavelength of 965-985 nm are being used for pumping Er:Glass lasers. Type SDL-6360 can give 1.0 W output power.

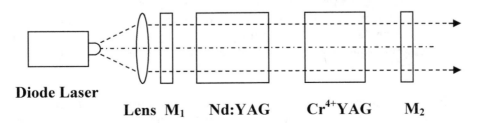

Diode Laser

Lens M₁   Nd:YAG   Cr^{4+}YAG   M₂

Figure-9: Diode Pumped Nd:YAG Mini-Laser

For giant pulse operation of Nd:YAG lasers, quasi continuous wave (QCW) laser diode array is used. SDL type 3200 gives an output of 100 W in pulse duration from 100 to 400μ second. For more pumping power stacked arrays can be used for pumping up to 4800 W [15]. Emission length of these arrays is 10 mm.

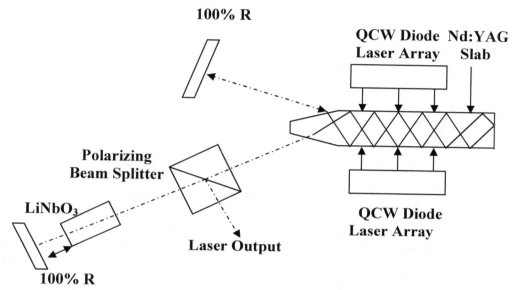

Figure 10-QCW Diode Pumped Nd:YAG Q-Switched Laser

The diode lasers have operating life of 10,000 Hrs in cw mode and 10^9 shots in pulse mode. Due to high cost and sacking problems of these arrays, still flash lamp pumped are used in spite of low operating life of 200 hours in cw mode and 10^7 shots in pulse operation mode.

5.3 EXCITATION / PUMPING OF GASEOUS LASERS

The Gaseous laser sources used for ranging purpose are He-Ne i.e. for distance measurement and TEA CO_2 lasers for Military range finders as discussed in chapter-3.

The He-Ne lasers are excited for population inversion either by D.C. electric discharge of excited by radio-frequency (rf) oscillator. In D.C. electrodes are parts of plasma tube such that electric discharge takes place through long narrow tube. To maintain electric discharge, depending upon length of plasma tube 2KV voltage is generally required for a plasma tube of 50 cm Length. Voltage multiplier is used which generate about 8 to 10 KV of voltage, which is applied to electrodes, and as soon as gas is ionized, voltage doublers takes over to maintain constant discharge current as described in chapter-7. Superimposing additional voltage in series can carry out intensity modulation of laser. Another method of excitation of He-Ne plasma tube is by radio frequency (rf) oscillator. A 50 W rf oscillator at rf of 20 to 30 MHz is used. This rf voltage is applied to copper foils wrapped round the length of plasma tube as shown in figure-11. He-Ne mixture ratio is 5:1 at pressure of 1.4 torr.

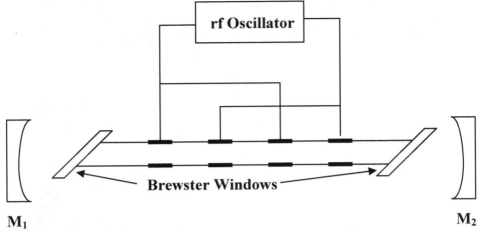

Figure 11- Radio Frequency Excited He-Ne Laser

The TEA CO_2 is operated in pulse mode. The gas filled is generally at atmospheric pressure and consists of mixture of He, CO_2 and N_2. The shape of electrode is Rogowski profile developed by Pearson and Lamberton (1972). The secondary electrodes, providing the pre ionization, consisted of a pair of fine Nichrome wires parallel to the main electrodes but offset from their center line and each connected to the cathode by a 30 pf coupling capacitor as shown in figure 12. Additional pre ionization can be provided by ultra violet source or radioactive source. The electrical excitation pulse is produced by discharging through a nitrogen-pressurized spark gap a low inductive capacitor of 35 nF, charged to between 15 and 25 KV. Due to radioactive source the discharge is without arcing. A pulse of 600 KW in pulse duration of 50 nano second at 1.06 μm is obtained [17].

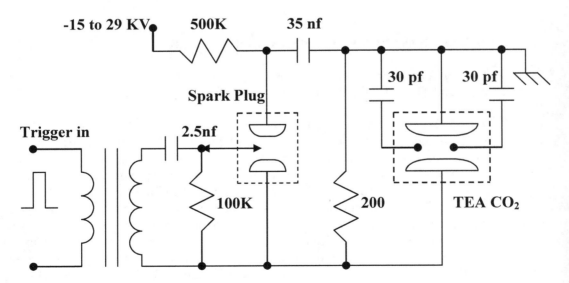

Figure 12 - Excitation of TEA CO_2 Laser

5.4 EXCITATION / PUMPING OF SEMICOUNDCTOR LASERS

Forward current through junction excites semiconductor lasers to create concentration of electrons and holes in depletion region or undoped region or quantum well region where they interact with confined radiation to generate stimulated emission. The continuous direct or pulse current should be above threshold. The intensity of these lasers depends upon forward current and can be easily modulated by modulating forward current. These lasers are used for measuring distance for survey purpose or pumping of solid-state lasers. To protect laser diode from reverse surge, a diode is connected across diode laser in reverse direction. A thermistor on heat sink of laser diode is used to operate laser diode at a temperature such that its output wavelength matches with absorption band of laser material. Temperature coefficient of wavelength of diode lasers is 0.27 to 0.3 nm / ^{0}C. The recommended operating temperature of these laser diodes is from –20 to 30 ^{0}C. The schematic of excitation of laser diodes is shown in figure-13. A monitoring photo diode on these lasers is used to control drive current in order to prevent the damage of these diodes from excessive radiation field.

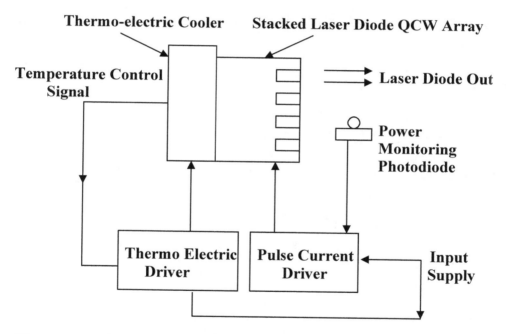

Figure 13 - Excitation of QCW Pump Laser Diode Stacked Array

SUMMARY

Xenon, krypton and semiconductor diode laser meets all requirements of pumping solid-state lasers. With diode pump lasers overall efficiency of solid-state lasers is 25%. But due to high cost of stripe arrays and their stacking problems, still xenon flash lamps are being user for pumping high power solid-state lasers required for ranging purpose.

Flash lamp in linear geometry with elliptical reflector offers most effective and efficient coupling of light output to laser rod. But diffuse reflector with $BaSO_4$ powder and cerium and samarium doped glasses in close coupling configuration offers efficient light output coupling to laser rods with better life as compared to silver or gold plated reflectors.

Gaseous lasers are excited with direct current or radio frequency voltage.

Semiconductor lasers are excited with forward current, continuous or pulsed.

REFERENCES

1. Data sheet, "Ruby Laser Rods", Crystal Product division of Union Carbide, USA.

2. Mansharamani, N., "Diode-Pumped Neodymium Laser's – Present Status and Future Prospects", Journal of Optics (India), Vol. 18, No.3, 1989, pp. 68 – 70.

3. Barnes, F. S., "Physical Characteristics of Xenon Flash Tubes", J. Soc. Motion Pictures and Televis. Engg. (USA), Vol.73, No. 7, September-October 1964, pp. 569.

4. Data Sheet F 1002 C – 2, "Linear Xenon Flash Tubes", M / S EG&G Electro – Optics, USA (1979).

5. Technical Bulletin 3, "An Overview of Flash Lamps and cw Arc Lamps", M/S ILC Technology Inc., USA.

6. Koechner, Walter, "Solid–State Laser Engineering", 3rd Edition, Springer-Verlog", Berlin (1992).

7. Final Report Vol. 1, "Optical Pumps for Lasers", 17 May1971- 16 July 1972, ILC Technology, Inc., USA.

8. Markiewicz,J.P. and Emmett,J.L., "Design of Flash Lamp Driving Circuit", IEEE J, Vol. QE-2, 1966, pp. 707.

9. Holzichter, J. F., Emmett, J.L., "Design and Analysis of a High Brightness Axial Flash Lamp" Appl. Opt. (USA), Vol. 8, No. 7, July 1969, pp. 1459.

10. Schultdt,S.B., Aagard,R.L., "An Analysis of Radiation Transfer by means of Elliptical Cylinder Reflectors, Appl. Opt. (USA), Vol. 2, No. 5, May 1963, pp. 509-513.

11. Edwards,J.G., "Some Factors Affecting the Pumping Efficiency of Optically Pumped Lasers", Appl. Opt. (USA), Vol. 6, No. 5, May 1967, pp.837-843.

12. Cann,D.M., "Optimal Reflectors for Coupling Cylindrical Sources and Targets of Finite Dimensions", Appl. Opt. (USA), Vol. 23, No. 4, 1984, pp. 601-606.

13. Jae-Kyung Hyung "A Study on the Improvement of Nd:YAG Laser Pump Cavities", New Physics, (Korean Phys. Soc.) Vol. 26, No.6, December 1986, pp. 456.

14. Data Sheet, "Single & Double Lamp Solid State Laser Pumping Chambers and Laser Cavity Filters", M/S Kigre, Inc., USA (1990).

15. Product Catlog 96/97, "Laser Diodes" M/S SDL Inc., USA.

128

16. Juyal, D.P., "Design of an rf Excited Helium – Neon Visible Laser and Study of the Optimal Condition for Gas Mixture and Pressure", Def. Sci. J, Vol. 22, October 1972, pp. 245-248.

17. Hammond C.R., Juyal, D.P., Thomas, G.C. and Zembrod, A, "Single Longitudinal Operation of Transversely Excited CO_2 Laser", J. of Physics E, Scientific Instruments 1974, Vol. 7 pp. 45 – 48.

18. Mansharamani, N. and Rampal, V.V., "Some Consideration for Optical Pumping of Solid State Lasers", Presented at Symposium on Science Technology of Infrared and Lasers, held at Defense Science Laboratory, Delhi from 2-3 February 1970.

CHAPTER – 6

Q – SWITCHING TECHNIQUES

6.1 INTRODUCTION

The quality factor Q of a cavity is defined as energy stored to energy lost per cycle i.e. decay of photon in the cavity. In a cavity with gain medium, the purpose of Q-Switching technique is to build a single giant pulse. In normal resonator with laser material, as soon as population inversion crosses threshold, laser output is obtained from partial mirror side of cavity. In this technique, the energy from the pump source is first stored in the laser material. When optimum population inversion in the laser material is obtained; introduction of cavity mirror or opening a shutter in the cavity forms a state of high quality factor Q from state of low Q, at the instant of optimum population. This results in release of stored energy in a single giant pulse.

McClung and Hellwarth [1] first demonstrated the Q-Switching in ruby laser using Electro-optic shutter in 1961. Since then lot of methods for generation of giant pulse have come up in last 40 years and have been classified under the following three heads.

Mechanical Methods.
Electro-optics Shutters.
Passive Q-switching Methods.

In mechanical methods spinning reflector i.e. rotating prism, frustrated total internal reflector or acousto-optics beam deflecting element in cavity is used for generation of giant laser pulse. These are slow switching techniques, simple, rotating reflector to form laser cavity, needs synchronization with flashing of lamp. This technique is used for low and medium gain laser materials like ruby, Nd:Glass and Er:Glass having large pulse build up time.

Electro-optics shutter comes under medium speed; can be used with any type of laser i.e. solid-state, gaseous or semiconductor; this technique is complex needs additional component in laser cavity with very good anti-reflection coating and also requires alignment with cavity mirrors. Advantage with this technique is that giant pulse can be generated at a precise time and cannot be generated with other techniques.

Passive Q-Switching has very fast switching speeds. In this technique, a reverse bleachable dye, color center crystal, black YAG in cavity, or semiconductor mirrors is used as Q-Switching element. In this technique, a sharp pulse is generated i.e. short duration pulse with high peak power. This method of switching cavity losses is automatic and does not need any synchronization with flash lamp firing. This method is mostly used with Nd:YAG laser material in the compact laser range finders.

Various type of Q-switching techniques and theory are discussed in [2, 3, 4]

6.2 THEORY

The quality factor Q of a cavity is defined as the ratio of energy stored in the resonator to power dissipated from the resonator per unit angular frequency ω_0

Where $\omega_0 = 2 \pi \nu_0 = 2 \pi / T_0$ ....[1]

If τ_c is average lifetime of photon in the resonator and $\Delta\nu$ is width at which the intensity of radiation falls half the maximum value of resonant frequency ν_0, Q is given by following relations

$$Q = \omega_0 \tau_c = \nu_0 / \Delta\nu$$ ....[2]

τ_c is related to the fractional power loss ε per round trip as

$$\varepsilon = t_R / \tau_c$$ ....[3]

Where $t_R = 2l' / c$ is the round-trip time of photon in a resonator having an free space optical length l'

The equation of lasers as denoted by population level in various states of laser material with three states and four states as a result of pumping rate W_p is given by

For three level lasers with energy state and population density as represented in figure 1a, rate equation is

$$\delta n_1 / \delta t = n c \phi \sigma + n_2 / n_1 - W_p n_1$$ ....[4]

Where $n = n_2 - g_1 n_1 / g_2$ ....[5]

And σ is stimulated emission cross section

or rate of change of photon density is given by

$$\delta\phi / \delta t = c \phi \sigma n - \phi / \tau_c - S$$ ....[6]

Where S is spontaneous emission.

For four level laser system energy state and population density as shown in figure-1b, the rate equation is

$$\delta n_2 / \delta t = - n_2 \, \sigma \, \phi \, c - n_2 / \tau_f + W_p \, (n_0 - n_2)$$ ....[7]

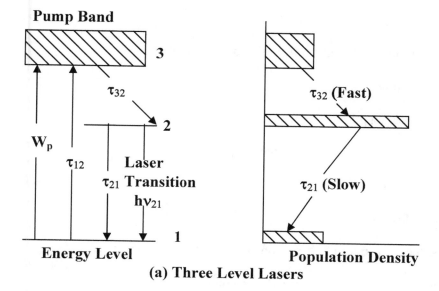

(a) Three Level Lasers

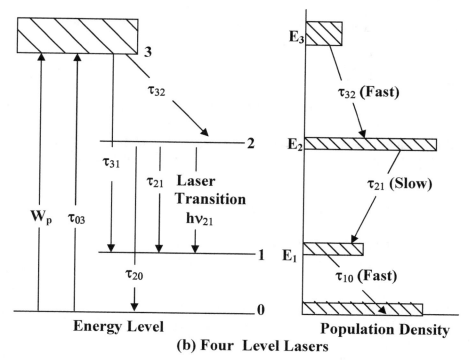

(b) Four Level Lasers

Figure-1 Energy Level and Population Density of Laser Material

Where fluorescence decay time τ_f of upper level is given by

$$1 / \tau_f = 1 / \tau_{21} + 1 / \tau_{20}$$ ....[8]

Effective radiation lifetime of metable-state is

$$\tau_{21} = A_{21}^{-1} \qquad \dots\dots\dots[9]$$

The photon built up rate is given from the rate equation as

$$\delta\phi / \delta t = \phi \ (c \ \sigma \ n \ l \ /l' - \epsilon / t_R) \qquad \dots\dots\dots[10]$$

Where l is optical length in laser material

The losses in a cavity can be represented as

$$\epsilon = - \ln R + L + \epsilon(t) \qquad \dots\dots\dots[11]$$

R is coupling losses of output mirror

L = losses due to scattering, diffraction and absorption in resonator

ϵ (t) = loss introduced by Q- switched element

The change in value of losses ϵ during Q-switch operation from ϵ_{max} to ϵ_{min} is represented as

$$\epsilon_{max} = - \ln R + L + \epsilon(t) \qquad \dots\dots\dots[12]$$

$$\epsilon_{min} = - \ln R + L \qquad \dots\dots\dots[13]$$

If n_i is initial population inversion before Q-switching and n_f is final population inversion after Q-switched pulse is built, energy E of Q-switched pulse is given by relation [3]

$$E = (h\nu \ A \ / 2\sigma \ \gamma) \ \ln (1/R) \ \ln (n_i / n_f) \qquad \dots\dots\dots[14]$$

Where A = beam cross section and

$$\gamma = 1 + g_2 / g_1 \qquad \dots\dots[15]$$

The pulse width of Q- switched pulse is given by

$$\Delta t_p = \tau_c \ \frac{n_i - n_t}{n_i - n_f [1 + \ln (n_i / n_t)]} \qquad \dots\dots[16]$$

Where n_t is population inversion at threshold represented by relation

$$n_t = [\ln (1 / R) + L] / 2\sigma L \qquad \dots\dots\dots[17]$$

6.2.1 Slow Q-Switching

In most of the range finders rotating prism is being used as Q-switched element. The loss rate variation is represented by

$$\varepsilon = -\ln R + L + B \cos (\omega t) \qquad \text{.........[18]}$$

Where ω is angular frequency of rotating prism.

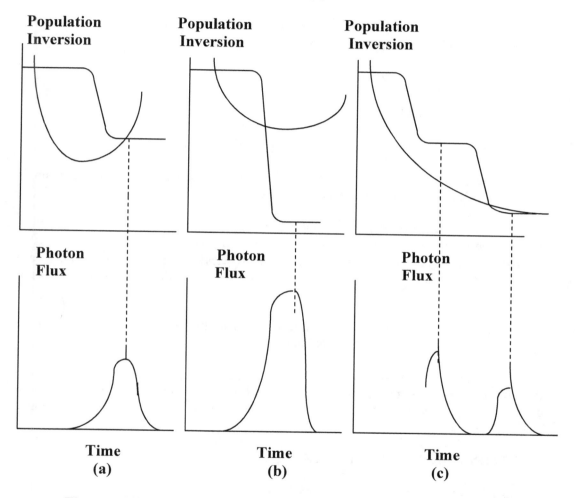

Figure-2 Pulse Build Up and Cavity losses for Slow Q-Switching

The photon flux starts to increase exponentially at $t = 0$. When photon flux is reached to maximum value ϕ_{max} i.e. at the time of the peak of laser pulse; the cavity losses should be minimum for optimum extraction of stored energy in laser material into a single giant pulse.

If g_0 is the gain in laser material, then the relation for radiation density built up is given by

$$\phi(t) = \phi_i \exp (2g_0 l - \varepsilon) \, t \, / \, t_R \qquad \qquad ......[19]$$

Where ϕ_i is initial photon density given by

$$\phi_i = n_2 . \Omega / 2\pi . \tau_c / \tau_f \qquad \qquad ....[20]$$

The build-up time t_D required for the evolution of the Q-switched pulse, let us say unto $\phi = \phi_{max} /20$ be given approximately as [4]

$$t_D = t_R \frac{\ln [\phi_{max} / 20 \, \phi_i]}{2g_0 l - \varepsilon} \qquad \qquad ....[21]$$

The pulse build-up time depends on gain and cavity length.

The pulse built and the cavity loss rate variation in case of rotating prism is illustrated in figure 2 i.e. for three extreme cases.

In figure-2a, the pulse delay time is longer than the switching time of the Q-switch. The photon flux starts to increase exponentially at $t = 0$, and at the time ϕ_{max} is reached i.e. pulse is emitted, Q-switched has already passed the point of ε_{min} . In this case pulse energy is less because the cavity losses are not at the time of maximum photon density.

In figure-2b, the pulse delay time is equal to switching time and in this case optimum energy is obtained in a single giant pulse.

In figure-2c, the Q-switch is much slower than the pulse delay time. This leads to the emission of several giant pulses. Therefore, for a given cavity length and gain or input energy, there is only one rotational speed of prism for which optimum energy in a pulse can be obtained.

6.3 MECHANICAL Q-SWITCHING

6.3.1 Rotating Prism

In this type of Q-switching [5, 6, 7], total internal reflecting (TIR) prism forms one reflector of laser cavity and it act either total reflector or partial reflector as shown in figure-3a and figure-3b. For very high power lasers configuration as shown in figure-3b is preferred as it can give high power single pulse as a result of enhanced speed due to multiple reflection from prism-1. The prism is rotated on a shaft (TIR edge parallel to shaft rotational direction) of D.C. permanent magnet motor with rotating speed of 18,000 revolution per minute (rpm) or A.C. synchronous motor operating on frequency of 400 cycles per second giving speed of 24,000 rpm to prism. The flashing of lamp is synchronized, so that laser cavity is formed, when all the optical energy from flash lamp is absorbed by laser material in creating a state of optimum population. The rotating speed of prism, in case mounted on shaft of D.C. permanent magnetic motor can be changed by varying voltage to match with pulse build up time for optimum extraction of stored energy or to avoid multiple pulses. Synchronization of prism position with flashing of lamp is achieved with as magnetic pick up in a coil placed close to magnet embedded in prism mount or with phase of A.C. voltage for synchronous motor. This type of Q-switching is used for ruby, Nd:Glass and Er:Glass lasers. For ruby laser roof of prism should be either parallel or perpendicular to the polarization of laser emission, so that reflected light remains plane polarized. In some cases, it has been used with Nd:YAG using multiple reflection [5, 6] as shown in figure-3b. TIR prism used has very high angle accuracy i.e. within 2 seconds for 90^0 as well as for pyramidal error for surfaces forming 90^0 angles. The quality of glass i.e. fused silica or borosilicate glass / BK-7 from which prism is fabricated should be of very high quality that is interferometeric grade.

Rotating prism devices are simple and inexpensive. They are insensitive to polarization, thermal and birefringence effects. But these devices are noisy and need frequent maintenance.

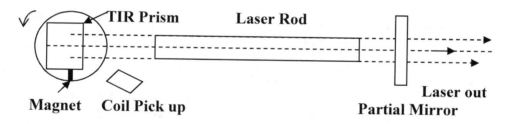

(a) Single Prism (Slow Switching Losses)

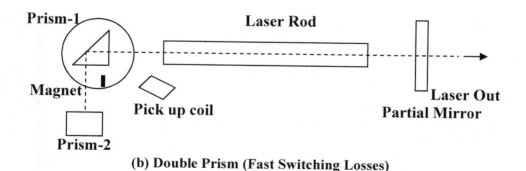

(b) Double Prism (Fast Switching Losses)

Figure-3 Rotating Prism Q-Switching Arrangements

6.3.2 Frustrated Total Internal Reflection

The light from TIR prism can be coupled out if another prism or optical plate as shown in figure-4, is placed parallel and very close to side face of prism, the coupled output light depends on distance d. Variation of reflectivity R of TIR prism to output coupled light T is given by relations [22, 23]

For polarization parallel to the plane of incidence,

$$\frac{R_p}{T_p} = \frac{(\mu^2 - 1)^4}{4\mu^2(\mu^2 - 2)} \sinh^2\left(2\pi\frac{d}{\lambda}\left\{\frac{\mu^2 - 2}{2}\right\}^{1/2}\right) \qquad \text{........[22]}$$

For polarization normal to the plane of incidence,

$$\frac{R_s}{T_s} = \frac{(\mu^2 - 1)^2}{\mu^2(\mu^2 - 2)} \sinh^2 2\pi\left(\frac{d}{\lambda}\left\{\frac{\mu^2 - 2}{2}\right\}^{1/2}\right) \qquad \text{............[23]}$$

It is assumed that T + R = 1   ......[24]

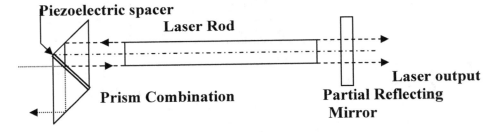

Figure-4 Q-Switching Using Total Internal Frustrated Reflection

The arrangement for Q-switching using frustrated internal reflection is shown in figure-4. Here the distance d between two prisms is almost zero, as the light is coupled out from total reflector, cavity losses are high and no lasing action can take place during pumping period. When all optical energy from flash lamp is absorbed in laser material in creating optimum population inversion, voltage on piezoelectric ceramic spacer increase gap d; thus prism becomes total reflector, as no more light is coupled out. Thus cavity losses are reduced and the energy stored in laser material is emitted as giant pulse through partial mirror. This type of arrangement can also be used in continuously pumped laser material for generating high repetition laser pulses. Unlike rotating prism Q-switching, this arrangement is free from noise; does not need frequent maintenance and the pulses can be generated at precise interval of time as requirement for precise coding in military systems. The details of this type of Q-Switching are given in references [2].

6.3.3 Acousto-Optic

In certain materials like quartz [13, 16] mechanical stain can be produced, if electric voltage is applied. These materials are known as piezoelectric material and the effect is known as piezoelectric effect. If alternate voltage is applied, vibrations can be produced in these materials. If their natural vibration frequency matches with frequency of alternate voltage, standing waves are generated in the material, which forms diffraction gratings. Thus light beam passing through material can be deflected depending upon the magnitude of this voltage. If this type of material is placed in laser cavity and voltage is applied, alignment get disturbed and lasing action cannot take place. Therefore during pumping, voltage is applied to piezoelectric crystal placed in laser cavity, alignment is disturbed and lasing action is prevented during pumping, when all light from flash lamp is pumped into laser material with state of optimum population inversion, at that instant voltage is switched off. As soon as voltage is switched off, mirror alignment is restored and all the energy stored in laser material is released in a single giant pulse. This type of Q-switching is illustrated in figure-5. The details of this type of Q-switching are given in references [11, 12, 13].

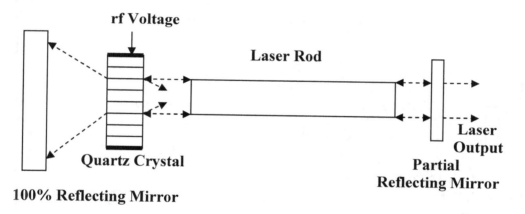

Figure-5 Acousto-Optic Q-Switching

6.4 ELECTRO-OPTIC Q-SWITCHING

In certain crystals or liquids, when electric voltage is applied in the direction of propagation of light (longitudinal field) or in transverse direction of propagation of light (transverse field) birefringence is produced in the medium i.e. velocity of light in orthogonal direction is different producing a phase difference. This result in change in direction of polarization that depends on electric field produced in the material and also travels length in case of transverse field. This effect is known as Kerr Effect in case of liquids and Pockel effect in case of solids. This device is called Kerr cell or Pockel cell. This device can act as shutter, if placed in combination with polarizer as shown in figure 6a and 6b. The magnitude of voltage should be such that it can change the direction of polarization by 45^0 or 90^0. This voltage is known as quarter or half wave voltage.

In this type of arrangement, laser action can be prevented during pumping and after all energy from lamp is pumped into material and optimum population is achieved, voltage on Pockel cell is switched off. Thus lasing produced results in production of giant pulse. The speed of Q-switching depends on how fast the voltage is switched off.

There is two type of Electro-optic effects which can be utilized in laser Q-switching. Pockel effect and Kerr effect, Pockel effect occurs in crystal, which lack a center of point symmetry and Kerr effect occur in certain liquids. Kerr cell normally not used as it needs 5 to 10 times more voltage than Pockel cell.

There are number of crystal which exhibit Electro-optic effect, but mostly Potassium di hydrogen phosphate (KDP) / Potassium di deuterium phosphate (KD*P) in longitudinal mode or Lithium Niobate (LiNbO$_3$) in transverse mode is used for Q-switching [17, 18]. In laser range finders LiNbO$_3$ is generally used as it has low temperature coefficient and does not need sealing from atmosphere. With anti reflection coating its transmission is 98%. Due to crystal symmetry, many Electro-optics coefficient in LiNbO$_3$ and KD*P crystals vanishes, only Electro-optic coefficient r_{22} for LiNbO$_3$ and r_{63} for KD*P contribute Electro-optic effect. The transverse voltage required to rotate direction of polarization of light of wavelength λ by 90^0 i.e. half wave voltage for LiNbO$_3$ is

$$V_{1/2} = \lambda d / 2 r_{22} \mu_0^3 l \qquad \qquad \ldots\ldots\ldots\ldots[25]$$

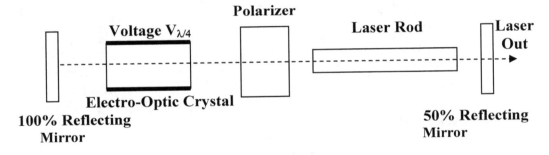

(a) Electro-Optic Crystal Operated at Quarter-Wave Voltage

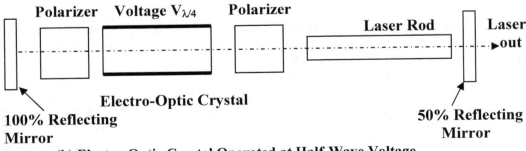

(b) Electro-Optic Crystal Operated at Half-Wave Voltage

Figure-6 Electro-Optic Q-Switching Arrangements

Where d is distance between electrodes and l is length of crystal and $\mu_0 = 2.237$ is ordinary refractive index of crystal.

$r_{22} = 5.61 \times 10^{-6} \ \mu m/V$

Half wave voltage for KD*P crystal operated in longitudinal mode is given by

$V_{1/2} = \lambda / 2 \ \mu_0^3 \ r_{63}$ ....[26]

$\mu_0 = 1.51$ is ordinary refractive index for KD*P

$r_{63} = 26.4 \times 10^{-6} \mu m/V$

6.4.1 Cavity Dumping

Electro-optics Q-switching can generate very sharp laser pulses of the order of few nano-second duration by a method known as cavity dumping. The cavity consists of 100% reflecting mirrors on both sides. Polarizing beam splitter along quarter wave plate is used in cavity beside Electro-optic crystal as shown in figure-7a. During pumping period no voltage is applied to Electro-optic crystal as shown in sequence of pulse operation of figure –7b. Due to quarter-wave plate in the cavity after polarizing beam splitter no lasing action can take place in the cavity i.e. after reflection from mirror-2 and again passing of this reflected light through quarter-wave plate this light is crossed inside the cavity and pumping energy from flash lamp is stored in laser material. As soon as all light from flash lamp is pumped into laser material in achieving optimum population inversion, a voltage pulse of magnitude of quarter-wave voltage is applied to Electro-optic crystal. Thus combination of quarter-wave plate and quarter-wave voltage, polarization of reflected light from mirror changes by 180^0. This reflected light now passes through polarizer into laser rod creating lasing action in the cavity. But laser

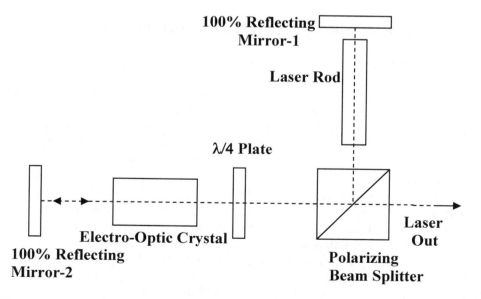

Figure-7a Electo-Optic Q-Switching Arrangement for Cavity Dumping

energy generated in cavity cannot escape through mirror, as both mirrors are 100%, thus whole laser energy is dumped inside the cavity. When maximum photon flux is in the cavity, quarter-wave voltage on Electro-optic crystal is switched off. For Nd:YAG laser duration of this pulse is 300 to 400 nano-seconds i.e. depends on cavity length and pulse build-up time of laser. This allows the flux to leak out of cavity through polarizing beam splitter as out put pulse. Duration of output pulse depends on fall time of quarter-wave voltage to Electro-optic crystal and cavity length. Generally duration of this laser pulse is less than 5 nano-second. This type of laser is used in bathymetry after second harmonic generation.

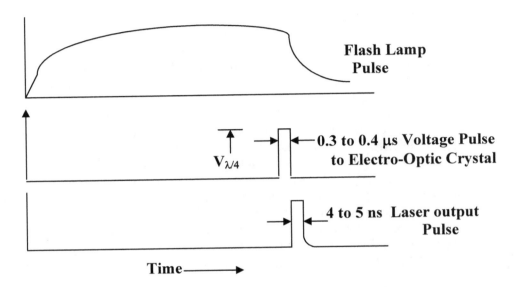

Figure-7b Sequence of Pulse in Cavity Dumping Operation

6.4.2   Damages due to Prelasing

In practice, due to clamping and inhomogeneities in the Electro-optic crystal, lensing effect of laser rod due to heat from pump source, non-perfect polarizer etc. The extinction ratio is limited to a few hundreds i.e. maximum to minimum transmitted intensity obtained between crossed polarizer as the applied voltage is varied through half wave voltage. If the gain of laser at this extension ratio, exceeds the loss, normal lasing action will occur before the instant of Q-switching, This phenomenon, which is termed as "prelasing" as Electro-optic crystal and polarizer combination does not act as perfect shutter. Pre lasing, generally occur just prior to the time of Q-switching, since the gain in laser material has maximum value at that instant. Prelasing occur in a small region of laser rod. Therefore with the opening of Electro-optic shutter, high power pulse is developed in this region, as compared to other region of laser medium. This high power density in this region damages optical component, Electro-optic crystal, and polarizer or laser rod. Therefore, before Q-switching, it should be ensured that no pre-lasing should take place in laser cavity. To avoid pre-lasing high grade component should be used, c-axis of Electro-optic crystal must be in the direction of laser action within angle of few arc of seconds and lensing effect in laser rod is compensated by additional optical component in laser cavity.

6.4.3 Procedure of c-Axis Alignment

Centering it on resonator axis, i.e. by producing and observing Maltese-cross pattern on resonator axis best performs alignment of c-axis. Illuminating the Electro-optic crystal with diffuse light source and observing the crystal between crossed polarizer on resonator axis i.e. laser rod axis. A pattern of cross-surrounded by a series of circle appears. The line connected the center of the cross should be adjusted on resonator axis. This ensures c-axis alignment with resonator axis. Further more to establish the proper condition for Q-switching crystal must be aligned so that its a-axis or b-axis is made parallel to the direction of polarization.

6.4.4 Post-Lasing

One or more pulses of lower amplitude may follow the main Q-switched pulse. These pulses occur after few hundred nanoseconds to tens of microseconds after the main pulse. The piezoelectric action of applied voltage compressed the Electro-optic crystal and when that voltage is removed, the crystal remains compressed for some time and takes some time to restore into unstained state. This compression generates a retardation of the optical wave by means of a stain birefringence effect, which creates loss in the cavity; this loss becomes smaller as the compression reduces which takes about 400 nanoseconds [18]. Due to reduction in this loss slower than pulse build up time another pulse is generated after main pulse because some energy remains in the rod after the output pulse. Thus efficiency of laser is reduced. This pre-lasing can be avoided by applying some reverse voltage to crystal and reducing fall time of flash pulse by pulse network in flash drive circuit. Oscillator amplifier combination can be used for very high power operation.

6.5 PASSIVE Q-SWITCHING

For passive Q- switching, elements used in Q- switching are

Reverse Bleachable Dye,
Lithium Fluoride Crystal with F_2^- Color Center (LiF:F_2^-),
Semiconductor Mirrors, and
Impurity Doped Crystalline Materials.

6.5.1 Reverse Bleachable Dye

Reverse Bleachable Dyes are used for ruby and Neodymium lasers. Placing of Dyes in laser cavity offers a most simple Q-switching technique, switching time is very fast; therefore these can be used with laser materials of very high gain in a cavity with small length. Unlike rotating prism Q-switching, no synchronization with flashing of lamp is required. The pulses produced are sharp with very small pulse duration of few nanoseconds. The Dye Q-switched laser is being used in Nd:YAG hand-held laser range finders. A dye switch consist of a glass cell of the order of 1 cm. Thickness placed between the laser rod and one of the mirror. Dye is filled in this cell with concentration such that optical transmission is reduced by 50% in one-way pass at laser wavelength. For Neodymium lasers, dye doped in acetate sheet is mostly used.

The dye crypto cyanine dissolved in methanol or several pathatocyanines dissolved in nitrobenzene (10^{-6} M solution) are employed as Q-switches for ruby [19, 20]. The concentration and the length of the absorption cell are so arranged that the cells transmit approximately 50%. The absorption lines correspond to electronic transitions of the phthatocyanine ring. The relaxation time of the phthatocyanine dye is of the order of 5×10^{-9} seconds, where as for cryptocyanine is of the order of 10^{-11} seconds.

Large varieties of passive Q-switch dyes [21] have been developed and are commercially available from M/s Eastman Organic chemicals of USA. Most popular dye for Q-switching Nd:YAG is bis [4-dimethylaminodibenzil(nickel)] in cellulose acetate sheet with optical density of 0.36 at 1.06 microns. When the dye is inserted into the laser cavity, it will look opaque to laser radiation until the photon flux is large enough to depopulate the ground level. If sufficient number of molecules are excited, the dye becomes complete transparent to the laser radiation. The saturation intensity I_s for saturable absorption can be

$$I_S = h\nu / \sigma_S \tau_S \qquad \qquad \dots \dots \dots \dots [27]$$

Where σ_S is the absorption cross section of dye molecules for laser radiation and τ_S is life time of excited state of dye molecules.

The structural formula of Bis[4-dimethylaminodithiobenzil (nickel)] is given below

Molecular Formula: $C_{32}H_{30}N_2NiS_4$
Molecular Weight: 629.55

This dye in cellulose acetate sheet has damage threshold is 300 megawatts per square centimeter. A compact laser source using Nd:YAG laser using Bis{4-dimethylaminodithiobenzil(Nickel)} in cellulose acetate sheet is shown in Figure-8.

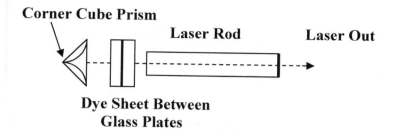

Corner Cube Prism

Laser Rod   **Laser Out**

Dye Sheet Between
Glass Plates

Figure-8 Passive Q-Switching Arrangement Using Reverse Bleachable
Dye in Cellulose Acetate Sheet.

6.5.2 Lithium Fluoride Crystal (LiF:F_2^-)

In LiF crystal lattice defect as shown in figure-9a i.e. color centers can be produced by excess lithium ions or by irradiating crystal with x-ray / γ–rays or by electron bombardment. A negative ion vacancy in a perfect lattice has the effect of an isolated positive charge. The LiF crystal can be grown easily and fabricated into a high optical quality. It is highly efficient, stable, compact and inexpensive for Nd:YAG laser Q-switched source as shown in figure-9b (absorption cross section of $2 \times 10^{-17} cm^2$ in near IR region) and can be operated at pulse repetition rate of 30 pulses per second with peak power of megawatt magnitude without any heat dissipation problems. The crystal life of several years and very high quality beam in fundamental mode and in nanosecond pulse can be generated for ranging application. The crystal has good thermal conductivity of

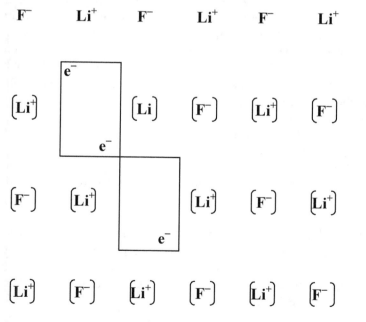

Figure-9b An F_2^- Center Consist of two Anion Vacancies with three Captured
Electrons

144

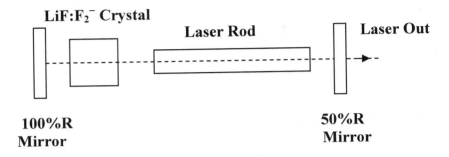

Figure-9b Passive Q-Switching Arrangement Using LiF:F₂⁻ Crystal as Q-Switching Element

2.7×10^{-2} cal/cm/s/^{0}C. With input energy of 10 joules output of 30mJ, 10-ns half width at half of maximum (FWHM) at pulse rate of 30 pps [24, 25].

6.5.3 Semiconductor Mirrors

A polished surface of Germanium (Ge), when used as a reflector, act as a passive Q-switch. The laser action increases the reflectivity of the semiconductor by inducing a high carrier concentration, and the output develops into a giant pulse. All the bulk semiconductor materials like silicon (Si), Ge, Indium Phosphide (InP), Indium Antimonide (InSb), Gallium Arsenic Phosphide ($GeAs_xP_{1-x}$), successfully operated as Q-switches for ruby laser, producing a mixture of normal and giant pulses, sometime single or multiple giant pulses. Most materials get damaged, if a single giant pulse has energy beyond 50 mJ. Therefore semiconductor mirror finds application in low energy, high repetition pate lasers. Now a days, multi quantum well semiconductor mirror known as semiconductor saturable mirror (SEMSAM) consisting of InGaAsP / InP are finding application in microchip lasers [32]. Unlike Cr^{4+}:YAG, SEMSAM can act as Q-switch elements for higher wavelength lasers. GaAs has also been used as passive Q-switch for Nd:YAG in inter-cavity doubling laser [26, 27].

6.5.4 Impurity Doped Crystalline Materials

Cr^{4+} doped impurities in garnet (YAG) crystal [28] and Co^{2+} in Magnesium aluminum oxide ($MgAl_2O_4$) [31] act as passive Q-switch for Nd:YAG and Er:Glass lasers respectively.

Cr^{4+}:YAG known as black YAG is at present best material for passive Q-switching lasers working in wavelength region from 0.9 to 1.1 µm and can also act as tunable laser in a wavelength region of 1.35 to 1.6µm. This material is very stable, rugged and efficient with long life as saturable absorber. It has produced billion of laser shot [29] without any appreciable deterioration.

In garnet crystal (YAG), Cr^{4+} state is present when YAG is also doped with Magnesium (Mg^{2+}) and calcium (Ca^{2+}). This kind of doping preserves the charge balance in crystals and their chemical composition is given by the stoichiometric formula

$$(Y^{3+}_{3-x}Mg^{2+}_{x}) (Al^{3+}_{5-x}Cr^{4+}_{x}) O^{2-}_{12}$$

During crystallization from the melt, a relatively small amount of Cr atoms are incorporated into the crystal as Cr^{4+} ions, and the remaining atoms as Cr^{3+}. Further oxidation of Cr^{3+} ions to Cr^{4+} ions should be performed in solid phase, by heating the crystal in the oxygen atmosphere. The crystal is grown by the Czochralski method in a r.f. heated iridium crucible and are pulled along [111] direction with the growth rate equal to 2 mm/h in nitrogen atmosphere with addition of about 3 vol. % of oxygen.

The crystal has been used as Q-switched element in producing high peak power, in lasers operated in billion shots without degradation in its properties [29]. In Nd:YAG micro laser it has produced pulses with duration of pico second as well as nanosecond duration [30, 31].

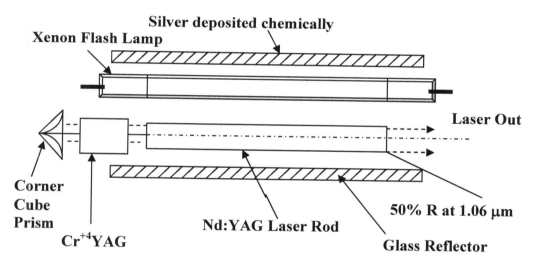

Figure-9c: Nd:YAG Q-Switched with Cr^{+4}YAG (IRDE)

A **Q-Switched Nd:YAG** Source as shown in Figure-9c is developed IRDE for use in medium repetition rate range finders using Cr^{4+} doped YAG with optical density 0.35 as Q-Switched element. With input energy of 18 Joules to xenon flash lamp to pump Nd:YAG laser rod of 5 mm diameter and 50 mm long. A corner cube prism forms as total reflector with 50% reflecting coating at the other end of laser rod. Laser output energy is 20 milli Joules in pulse width of 17 ns. Laser operates at pulse repetition rate of 4 pulses per second using closed forced air-cooling for flash lamp and laser rod.

A saturable absorber Co^{2+}:Mg Al_2O_4 crystal has been recently used as passive Q-switch element for Nd:YAlO_3 operating at 1.34 μm and Er:Glass laser at 1.54 μm [32] producing pulses of 45 nanosecond duration with energy of 2.5 mJ.

SUMMARY

A giant laser pulse is required for laser range finders for military or diffuse reflecting type of targets. Various Q- switching techniques for obtaining giant pulses are used in range finders for various roles. Rotating prism Q-switching is simple and cheap technique for obtaining giant pulse, but disadvantage with this technique is that for optimum extraction of stored energy in a single pulse from excited laser material, switching time has to match with pulse build up time. With high gain laser material like Nd:YAG, better and cheapest technique for giant pulse generation is passive Q-switching technique; disadvantage is that pulse cannot be generated at precise interval of time, i.e. for time pulse coded systems required in military devices. Precise time control for generation of pulse is possible with Electro-optics, acousto-optics and frustrated total internal reflection techniques. These techniques can be adapted to laser operating at any wavelength. Optimum extraction of energy and control on pulse width is possible with cavity dumping using Electro-optic shutter in laser cavity. Semiconductor saturable absorber mirror can be used as passive Q-switch element for lasers operating in mid infrared region.

LIST of SYMBOLS USED

A = Laser beam cross section

A_{21} = Spontaneous emission coefficient

B_{21} = Stimulated emission coefficient

B_{12} = Absorption coefficient

c = Velocity of light in vacuum

d = Distance between TIR prism for frustrated total internal reflection or distance between electrodes of electro optic crystal.

E = Energy of Q-switched pulse in Joules.

g_1, g_2 = Degeneracy of 1,2 energy state

g_0 = Gain of laser material due to population inversion

l = Optical length in laser material or electro optics crystal l' = Free space optical length in resonator

L = Losses in resonator due to scattering, diffraction and output light coupling

n_o = Population in ground state due to population inversion

n_1, n_2 = Population density state in 1, 2 energy levels

n_i = Initial population density due to population inversion

n_f = Final population inversion after Q-switched pulse

R = Coupling loss of output resonator mirror

R_p/R_s = Reflectivity of light in TIR prisms combination for frustrated total internal reflection for light with plane of polarization parallel / perpendicular to plane of incidence

r_{22} , r_{63} = Nonlinear coefficient of Electro optic crystal

Q = Quality factor of laser cavity

S = Spontaneous emission

T_0 = Time period of resonant frequency

T_p / T_s = Output coupled light from TIR prism combination for frustrated total internal reflection for light with plane of polarization parallel / perpendicular to plane of incidence

t = Time

t_D = Pulse build up time

t_R = Round trip time of photon in a resonator

$V_{1/2}/V_{1/4}$ = Value of voltage applied to Electro-optic crystal to produce 90^0 / 45^0 phase difference between ordinary and extra index light

W_p = Pump rate

ε = Loss per round trip of light in laser cavity

ε (t) = Losses introduced in laser cavity by Q-switch element

ε_{max} and ε_{min} = Maximum and minimum loss in laser cavity as a result of Q-switching

Δt_p = Pulse width of Q-switched pulse

ϕ = Photon flux density in laser resonator

ϕ_i = Photon density in laser cavity /resonator before Q-switching

ϕ_{max} = Photon density in cavity at the instant of peak of Q-switched pulse

λ = Wavelength of light

ν_0 = Resonant frequency of light in laser cavity in cycles per seconds

$\Delta\nu$ = Frequency width in laser resonator at which intensity of radiation falls half the value of light intensity at resonant frequency

Ω = Laser beam divergence

σ = Stimulated emission cross section

τ_f = Effective fluorescence decay time of metastable state

τ_{21} = Life time of metastable state

μ = Refractive index

μ_o = Refractive index for ordinary light

ω_0 = Angular resonant frequency in radians per second

REFERENCES

1. Hellwarth, R.W., Advances in Quantum Electronics, Columbia University Press, New York, pp.334 (1961).

2. Steele, L. Earl, "Optical Lasers in Electronics", John Wiley and Sons Inc., New York, (1968).

3. Bhattacharyya, A.N., Rampal, V.V. and Mansharamani, N., "Q-Switching Techniques for Solid State Lasers Operating at Room Temperature", J. of Scientific & Industrial Research, Vol. 27, No. 10, October 1968, pp. 380-385.

4. Wagner, W.G. and Lengel, B.A., "Evaluation of the Giant Pulse in Lasers", J. Appl. Phys., Vol. 34, No.   , 1963, pp.2040-2046.

5. Mansharamani, N., "Q-Switching Techniques for Portable Laser Range Finders", Presented at Symposium on Infrared and Instruments, held at Bhaba Atomic Research Center, Bombay in April 1980.

6. Arecchi, F.T., Potenza, G. and Sona. A., "Transient Phenomena in Q-Switched Lasers: Experimental and Theoretical Analysis", IL NUOVO CIMENTO Vol.XXXIV, No. 6, 16 December 1964, pp.1449-1472.

7. Benson, R.C. and Mirarchi, M.R., the Spinning Reflector Techniques for Ruby Laser Pulse Control", IEEE Transactions on Military Electronics, Vol. 8, No.1, January 1964, pp. 13-21.

8. Chernoch, J.P. and Titld, K.F., "Performance of Multi pulse, Multi prism Q-Switch", Proc. IEEE, Vol. 52, No. 7, July 1964, pp. 859-860.

9. Delay, R. and Sims, S.D., "An Improvement Method of Mechanical Q-Switching Using Total Internal Reflection"" Appl. Optics (USA) Vol. 3, No. 9, September 1964, pp.1063-1066.

10. Asthemen, R.W., el ta, "Infrared Modulation by means of Frustrated Total Internal Reflectors", Appl. Optics (USA), Vol. 5, No.1, January 1966, pp.87-91.

11. DeMaria, A.J., "Ultrasonic-Diffraction Shutter for Optical Laser Oscillator", J. Appl. Phys., Vol. 34, No. . 16 October 1963, pp. 2984-2988.

12. Supryrowiez, V.A., Giant Laser Pulse Formation using Ultrasonic Q-Spoiling", J. Appl. Phys., Vol.37, No. , February 1966, pp.778-784.

13. Naoya Uchida and Nobukazu Niizeki, "Acousto optic Deflection Materials and Techniques", Proc. IEEE, Vol. 61, No. 8, August 1973, pp.1073-1092.

14. Cheslu, R.B., Karr, M.A. and Geusic, J.E., "An Experimental and Theoretical Study of High Repetition Rate Q-Switched Nd.:YAG Lasers", Proc. IEEE, Vol. 58, No. 12, December 1970, pp. 1899- .

15. Gordon, E.I., "A Review of Acoustooptical Deflector and Modulation Devices", Proc. IEEE, Vol. 54, No. 10, October 1966, pp. 1391- .

16. Dixon, R.W., "Photoelastic Properties of Selected Materials and their Relevance for Applications to Acoustic Light Modulators and Scanners", J. Appl. Phys., Vol. 38, No. 12, December 1967, pp. 3149-3153.

17. Davies, M.B., Sarkies, R.H. and Wright, J.K., "Operation of a Lithium Niobate Electro-Optic Q-Switch at 1.06 μ", IEEE J. of Quantum Electronics, Vol. QE-4, No. 9, September 1968, pp. 533-535.

18. Hilbay, R.P. and Hoop, W.R., "Transient Elosto Optic Effects and Q-Switching Performance in Lithium Niobate and KD*P Pockel Cell", Appl. Optics, Vol. 9, No. 8, pp.1939-1940 (August, 1970).

19. Master, J.I. and Murry, E.M.E., "Comparison of Passive Q-Switch Components and Observation of Scattering Effects", Proc. IEEE, Vol. 53, No. 1, January 1965, pp. 76- .

20. Soffer, D.H. and Hoskins, R.H., "Generation of Giant Pulse from a Nd Laser by a Reversible Bleachable Absorber", Nature, 17 October 1964, Vol. 204, No. , pp. 276-.

21. Product Catalogue, "Q-Switch Dyes", Eastman Organic Chemicals", (USA).

22. Szalo, A. and sleir, R.A.S., "Theory of Laser Giant Pulsating by a Saturable Absorber", J. Appl. Phys., Vol. 36, No. 5, May 1965, pp. 1562-1566.

23. Skeen C.H. and York. C.M., "The Operation of a Neodymium Glass Laser Using a Saturable Liquid Q-Switch", Appl. Optics, Vol. 5, No. 9, September 1966, pp. 1463-1464.

24. Junewen Chen, Fu, I.K. and Lee, S.P., "LiF:F_2^- as a High Repetition Rate Nd:YAG Laser Passive Modulator", Appl. Optics (USA), Vol.29, No. 18, 20 June 1990, pp. 2669-2674.

25. Demchuk, M.T., Mikkailuv and Kuleshev, "Saturable Absorber Based on Impurity and Defect Centers in Crystals", IEEE J. of Quantum Electronics, Vol. 31, No. , 1995, pp. 1738-1741.

26. Charmichad, C.H.and Supsor, G.N., "Generation of Giant Maser Pulse Using a Semiconductor Mirror", Nature, Vol. 202, No. , 23 May 1964. pp. 787-788.

27. Sooj, W.R., Geller, M. and Borfeld, D.P., "Switch of Semiconductor by a Giant Pulsing by a Saturable Absorber", J. Appl. Phys., Vol. 36, May 1965, pp.1562- .

28. Frukach, Z., Lukasiewicz, T., Malinowski, M. and Mierczyk, Z. "Cr^{4+}:YAG Crystal Growth and its Optical Properties", Technical Report, Military University of Technology, ul. Kaliskiego 2, 00-908, Warsaw, Poland.

29. Stephen, M. A., Dallas J.L. and Afzal, R.S., "Multi-Billion Shots, High Influence Exposure of Cr^{4+}:YAG Passive Q-Switch", Proc. SPIE, Vol. 3244, 1998, pp. 517-521.

30. Shimony, Y., Burshtein, Z. and Kalisky, Y., "Cr^{4+} :YAG as Passive Q-Switch and Brewster plate in a Pulsed Nd:YAG Laser", Vol. 31, No. , 1995, pp. 1738-1741.

31. Agnesi, A., Dell' Acqua, S., Morello, C., Piccinno, G., Reali, G.C. and Zhaoyang Sun, "Diode- Pumped Neodymium Lasers Repetitively Q-Switched by Cr^{4+}:YAG Solid- State Saturable Absorbers", IEEE J. of Selected Topics in Quantum Electronics, Vol. 3, No. 1, February 1997. pp.

32. Ursula Keller, Kurt J. Weingarten, Franz X. Kartnu, Daniel Koft, Bernal Brown, Isabella D.Tuz, Regula Fluck, Chemens Honninjer, Nicolin Matuschek and Juers Ausder Au, "Semiconductor Saturable Absorber Mirror Nanosecond Pulse Generation in Solid State Lasers", IEEE J. on Selected Topics in Quantum Electronics, Vol. 2, No.3, September, pp. 435-453.

33. Birnbaun,M. and Stocker,T.L., "Giant Pulse Laser Operation with Semiconductor Mirrors", IEEE J. of Quantum Electronics; Vol. QE-2, July 1966, pp. 184-185.

34. Basu,M.K., Bhatnagar,V.S., Mansharamani,N. and Rampal,V.V., "Performance Optimization of an Indigenous Q-Switched Nd:Glass Laser", Jr. of Instrument Society of India, Vol. 7, No. 1&2, 1977, pp. 23-28.

CHAPTER-7

ELECTRONICS

7.1 INTRODUCTION

Electronics play a very important part for reliable design of laser range finders. Starting from charging of energy storage condenser to range measurement, electronics requirement for ranging techniques is described under following heads

Laser Power Supply
Pulse Driving Circuit for GaAs Laser
Trigger Circuits
Q - Switching Circuits
Range Measuring Electronic
Distance Measurements
Low Light Level Detection Circuits
Safety Circuits

7.2 LASER POWER SUPPLY

The purpose of laser power supply is to generate high voltage to (a) charge energy storage condenser for flash lamp driving circuit or pulse discharge circuit for TEA CO_2 laser; (b) high voltage to continuous discharge of He-Ne laser at low current; (c) a high voltage and very low current for trigger circuit or for Pockel cell. A +5 volt supply is required for TTL IC-circuits in range measuring unit. A low noise with proper de-coupling circuit + 12 V power supply is required for preamplifier and post-amplifier in low level detection circuit. Therefore various types of power supply circuits are described below

7.2.1 Power Supply for Flash Lamp Operation:

The flash lamp for pumping solid state lasers; energy storage condenser is to be charged between 1 to 2 kilo-volts (KV). For this purpose various power supply configurations [1] have been used, but most simple and efficient charging system is flyback high frequency inverter with operating frequency of 20 to 40 kilohertz (KHz). Advantages with this type of inverter are

*No series element is required to limit current between supply and condenser.

*Few components are required in the unit i.e. a high voltage fast switching transistor, fast recovery high voltage diode and a high frequency ferrite, preferably in pot core shape

152

*Turn ratio between primary and secondary; unlike other type of converter on high output voltage, but ratio depend upon the breakdown voltage of transistor i.e. between collector to emitter.

*For compact design, it can charge a high voltage electrolytic or metallised capacitor.

*Main power dissipation is in transistor and high voltage diode, therefore with proper choice of transistor with fast turn off time and very low saturation voltage between collector to emitter and fast recovery rectifying diode, 95% overall efficiency can be achieved.

*It provides fast charging to condenser and therefore can be used in high repetition rate laser range finding systems.

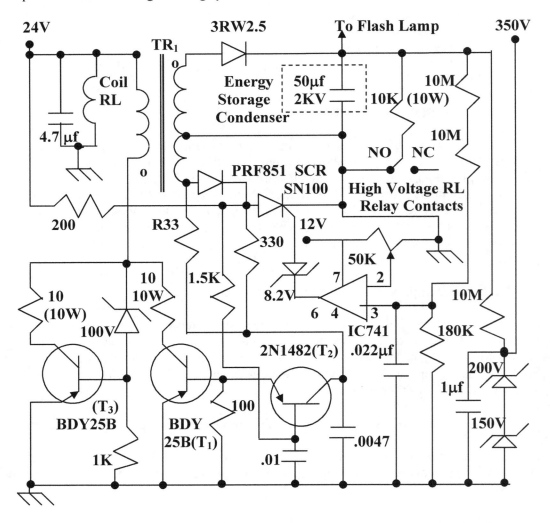

Figure-1a Fly back Converter for Charging Metallize Energy Storage Condenser.

Two type of high frequency flyback converter developed at IRDE for xenon flash lamp operation for pumping ruby, Nd:Glass. Er:Glass or Nd:YAG are illustrated in figure-1a and 1b, The circuit diagram of inverter shown in figure-1a can be used to charge polyester or metallised polyester condenser unto 2 KV, while shown in figure-1b can charge one electrolyte condenser (rating 450 V) in series by charging one condenser to a value of + 450V and other to − 450 V i.e. maximum voltage of 900 V is available for flash lamp operation. As soon as the battery voltage is applied to the unit,

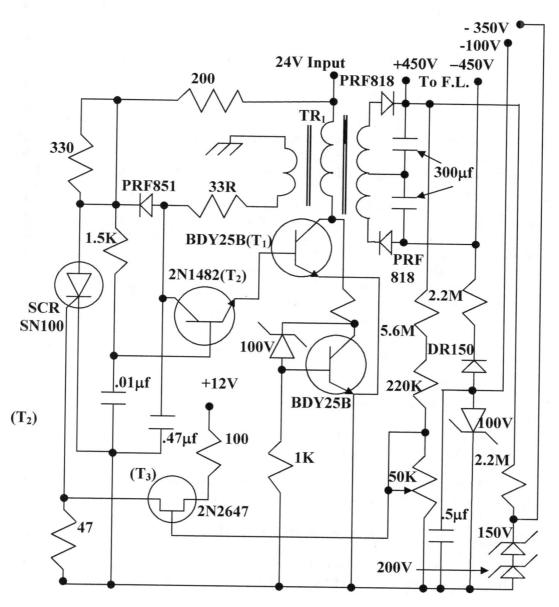

Figure-1b Power Supply for Charging Electrolytic Energy Storage Condenser for Maximum voltage of +450 and −450V for Flash Lamp.

transistor T_1 starts conducting by means of base drive through a resistor R (33 Ω) and current in the primary of transformer T_{R1} increases slowly depending on the inductance of primary winding. This increasing flux induce the voltage in base drive winding of transformer to maintain the base drive through resistor R and transistor T_2 as the silicon controlled rectifier's (SCR) anode voltage in voltage sensing circuit is high. The current in transistor T_1 increase to a point where the current gain β of transistor does not increase further, but start falling with current. As the current does not change in the primary of T_{R1}, the base drive voltage is reversed suddenly, switching off the transistor T_1. Since the energy stored in the inductance of primary of transformer is ½ L I_{max}^2 can not flow in base or collector of transistor to ground path, but gets favorable condition to charge energy storage condenser C through a fast recovery diode D to a voltage.

$$V = (L / C)^{1/2} I_{max} \qquad \qquad \dots\dots\dots[1]$$

This process is repeated as again the transistor gets drive through base drive resistor and the voltage on energy storage condenser increases to a value, sensed by unijunction transistor T_2 whose emitter is connected to voltage dividing resistors connected to energy storage condenser. The value of the resistor is adjusted to a required value of energy to be stored on condenser i.e. required for laser operation. As soon as the voltage reaches to a value of 8 volts at base of unijunction transistor, a voltage pulse at its emitter drives SCR to conduction. Thus switching off the converter, as no base drive voltage is available to transistor T_1 since base gets voltage, when SCR does not conduct and its anode is at high potential. Similarly circuit shown in figure-1a works and charges energy storage condenser to any voltage depending upon setting of potentiometer, in this case voltage comparartor is used instead of unijunction transistor to switch off the converter as soon as required energy is stored on energy storage condenser. In this circuit negative voltage is not available for reference photodiode.

Beside, charging energy storage condenser, unit has following additional features.

*350 V is available for shunt trigger circuit of flash lamp.

*The converter is switched off automatically, as soon as the condenser receive require energy and a visible indication is obtained.

*Energy storage condenser is kept shorted during off period of supply. This requirement is there from point of view of personnel safety.

* −100v is available for transmitted pulse reference detector circuit.

*In case of any failure in voltage sensing /cut out circuit, the condenser voltage is limited; voltage of transistor is clamped through 100V zener diode and transistor T_3.

7.2.2 Power Supply for TEA CO$_2$ Laser

Power supply for excitation / pumping of TEA CO$_2$ consist of push pull oscillator with primary to secondary winding turn ratio of about 1:100 as shown in figure 2. The oscillator works on any voltage in between 12 to 24 V DC supply. The voltage is supplied to oscillator at a fixed voltage through a series voltage regulator. The A.C. stepped up voltage on secondary of transformer is further stepped up to 25 KV D.C. by means of voltage multiplier circuit consisting of 13 diodes and 13 capacitor to charge a condenser of 150 nano farad to a maximum of 25 KV D.C. This converter also charges a 0.68 μf capacitor to a voltage of 300-350 V; this capacitor is discharge through a SCR and primary of trigger coil to initiate ionization at 30 KV through auxiliary wire electrode in TEA CO$_2$ laser tube. As soon as ionization in TEA CO$_2$ tube is produced, main condenser of 150-nano farad is a discharge exciting CO$_2$ molecule for laser action [2].

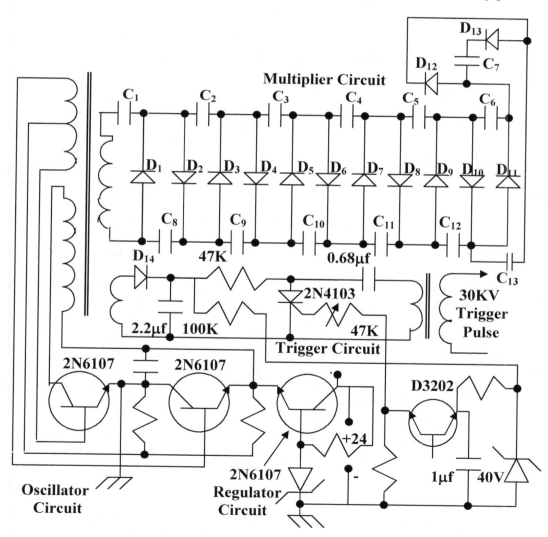

Figure-2: Power Supply for TEA CO$_2$ Laser

7.2.3 Power Supply for He-Ne Laser

He-Ne laser is generally used for distance measuring equipment for survey purpose due to their high directivity. The photo multiplier tube used in laser receiver for detection of light at wavelength 6328 A^0 in visible region is very sensitive and fast. A DC power supply with operating voltage of 2 KV and tube needs trigger voltage of 8 to 10KV to excite the plasma tube. The power supply developed at IRDE is shown in figure-3. The beam can be intensity modulated by either varying its drive current or use of Electro-optic modulator at the beam-transmitting end. The input voltage from inverter or AC main is stepped up to 1KV before applying to voltage multiplier circuit. As soon as plasma tube is connected to supply, multiplier circuit excites plasma tube and subsequently voltage doublers supplies current to maintain discharge.

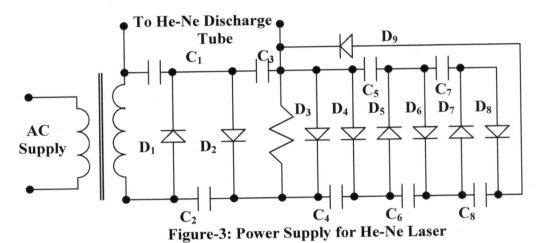

Figure-3: Power Supply for He-Ne Laser

7.2.4 Power Supply for Trigger Circuit / Electro-Optic Q-Switch

Power requirements for trigger circuits and Electro-optic switch are very small, except simmer operation of flash lamp. Generally power requirements for shunt or series injection trigger circuit in range finder is derived from fly back inverter by charging condenser of 0.5 µf charged to voltage of 350V for shunt trigger circuit and 0.1µf to be charged to voltage of 1KV for series injection trigger circuit. The Electro-optic switch require a voltage of 1.6 or 3.2 KV for Lithium Neobate crystal of 25 mm long and 9 mm transverse electrode distance for quarter wave or half wave voltage operation. This is met with fly back inverter charging energy storage condenser with additional winding to charge condenser of 0.1 µf through another diode. In case of simmer operation of flash lamp additional power supply of 800V is used for simmer operation. This supply is stepped up to half wave or quarter wave with voltage multiplier for Electro-optic switch.

7.2.5 Low Voltage Power Supply

Low voltage supplies required for laser range finder are +5V and +12V for TTL integrated circuits and preamplifier and amplifier circuits in laser receiver. If input supply is +24 volts, switching regulator for +5V is used, as current is about 1 Ampere (A) and for +12V supply series regulator is used as current requirement is less than 0.15A.

7.2.5.1 +5V Switching Regulator Circuit

The circuit diagram of +5V switching regulator with maximum load current of 1A, short circuit protected and over voltage protection at 5.2V is shown in figure 4. Input voltage of +24V is fed to this circuit through electro magnetic interference (EMI) filter.

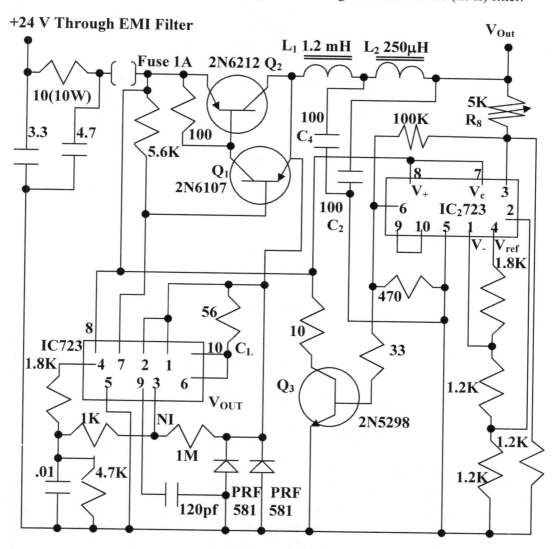

Figure-4: +5V Switching Regulator with Over Voltage Protection

The voltage is stored on the condenser C_4 through a ferrite inductor L_1 by the switching action of transistor Q_2 through voltage comparator in IC_1 and transistor Q_1. The switching frequency is about 15KHz for a load of 1A at 24V input. The ripple voltage on C_4 is filtered through L_2 and C_2. The over voltage protection at 5.2V is adjusted by variable resistor R_8 connected to voltage comparator in integrated IC_2 which in turn make Q_3 conducting and blowing off the fuse F.

7.2.5.2 +12V Series Regulator

The circuit diagram of +12V regulator is shown in figure 5. It consists of a hybrid voltage regulator IC µA78HG, with variable resistor for adjusting output voltage. The feed through capacitors C_1, C_1 at input and C_4, C_4 at output works as decoupling capacitors for transit noise. This voltage regulator is housed in a box and output voltage is fed to amplifier and preamplifier through r.f. coaxial cable.

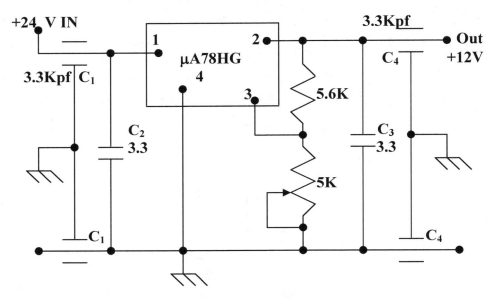

Figure-5: +12V Voltage Regulator

7.3 PULSE DRIVING CIRCUIT for GaAs LASERS

Semiconductor lasers are operated in forward voltage bias mode. Therefore, power supply requirements are low voltage and high current depending upon threshold current of semiconductor laser and output power requirements. For pulse operation, a fast transistor or silicon-controlled rectifier (SCR) is used to discharge a condenser of small value through laser diode in forward voltage bias mode with a fast diode connected across laser diode for reverse surge protection. As the threshold current and output power of laser diode depends upon its operating temperature, value of operating current should vary with temperature for constant output power of laser diode. Now a days, laser diodes are available with a power monitoring photodiode on the same chip. If the photodiode is biased in conduction mode, its resistance will vary with input optical power, if this photodiode is used to bias transistor supplying current to laser diode in such a way as shown in figure-6a, that diode drive current decrease if the resistance of photodiode decreases. Thus with monitoring photodiode constant laser power can be achieved at any temperature value. Modulating the input drive current to laser diode can modulate the output of laser. The sinusoidal modulation of laser diode output at various frequencies is required to measure distance by comparing phase of reflected laser with outgoing laser.

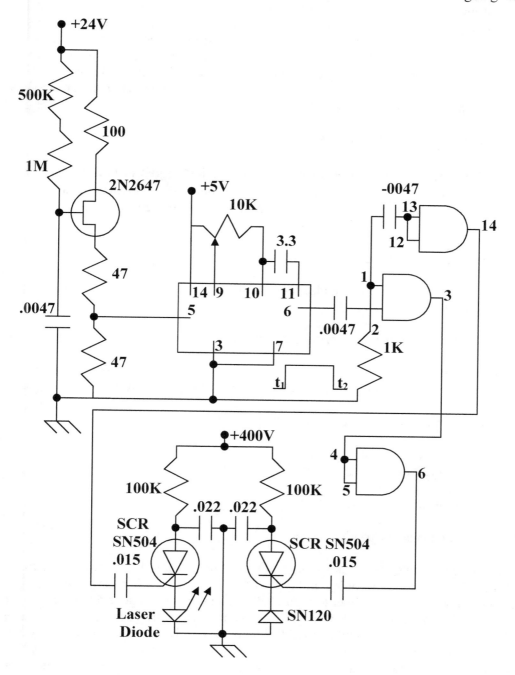

Figure-6: Double Pulse Coded GaAs Laser

160

Circuit diagram of laser diode drive developed at IRDE [3] for its operation in double pulse mode is shown in figure-6b. The circuit of unijunction transistor Q1 generates pulses of required pulse repetition frequency. Each pulse of the transistor triggers a monostable IC modulator 54121. The output pulse of this multivibrator has a pulse width $(t_2 - t_1)$ set by requirement of desired separation between laser pulses. Differentiating this multivibrator output pulse gives two pulses at time t_1 and t_2, which then triggers two SCR for obtaining the respective driving current pulses for laser diode. A capacitor charged to voltage of 400V discharge through laser diode by firing SCR. The pulse width and output power depends upon capacity of condenser and voltage it is charged. In this case pulse duration is 100 nano second.

7.4 TRIGGER CIRCUITS

A trigger circuit is used to produce ionization in gas to reduce operating voltage, so that discharge in flash lamp or gas laser plasma tube can take place at low voltage. Life, operating efficiency of flash lamp and gas laser depends upon the trigger methods [4] used.

The requirements of trigger circuit are

*Production of uniform ionization, axially between electrode and not near the walls of linear lamp.

*Sufficient trigger energy and operating voltage as recommended by manufacturer.

*Trigger circuit should be compacts and must be kept close to lamp.

The circuits used for triggering flash lamps are

Shunt Trigger Circuit
Series Injection Trigger Circuit.
Simmer Operation.
Pseudo-simmer Operation.

7.4.1 Shunt Trigger Circuit

The shunt trigger circuits are used in hand-held / compact laser range finders, as the trigger transformer is a ringing coil type and can be fabricated in very compact size. The shunt trigger circuit is shown in figure 7. The condenser C_1 is generally charged to a voltage of 350V and is discharge through SCR and primary of trigger coil T. The voltage is stepped up to a value between 15 to 20KV by trigger coils. For this turn ratio between primary and secondary of trigger coil is 1:50. The peak discharge as given by relation below should not exceed maximum rated anode current of SCR. If C is the value of condenser charged to a voltage V and discharge through primary of trigger coil having L_p as inductance, then peak discharge current I_p is given by

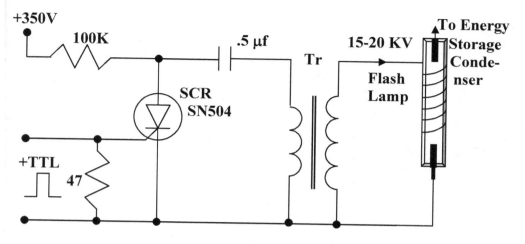

Figure-7 Shunt Trigger Circuit for Flash Lamp.

$$\tfrac{1}{2} C V^2 = \tfrac{1}{2} L_p I_p^2$$

$$I_p = (C / L_p)^{1/2} \cdot V \qquad \qquad \ldots\ldots\ldots[2]$$

For better coupling between primary and secondary coil, a small ferrite is placed coaxially at center of coil. Further, discharge capacitor C should be such that series resonant frequency with primary inductance should be equal to ringing frequency of secondary coil such that

$$C \cdot L_p = C_s \cdot L_s \qquad \qquad \ldots\ldots\ldots[3]$$

Where C_s is distributed capacitance of secondary coil.

The ringing frequency is generally 100KC/s. The fall in amplitude of ringing pulse is 10% at 10th cycle measured with capacitance probe.

7.4.2 Series Injection Trigger Circuit

The series injection circuit is shown in figure-8. It consists of high voltage pulse transformer T_r with thick wire secondary having turn ratio between primary and secondary as 1:20. The primary and secondary windings are wound on high frequency ferrite core with pot or torrid shape. A condenser of 0.1 µf charged to a voltage of about 1000V is discharge through primary of SCR. Thus momentarily a high voltage developed at secondary along with voltage of energy storage condenser appears across flash lamp, produces ionization and discharge of main energy storage condenser through lamp and secondary of transformer T_r. The air inductance of secondary should match the impedance of lamp for critical damping. As the transformer is bulky, this type of trigger circuit is not used with portable laser range finders, though 10% more efficiency and better triggering reliability is achieved as compared to shunt triggering method.

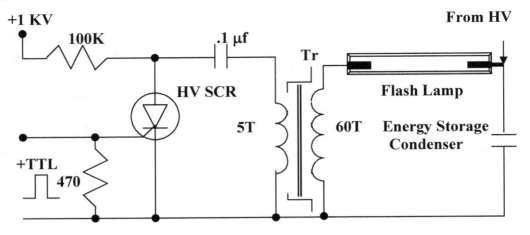

Figure-8: Series Injection Trigger Circuit for Flash Lamp.

7.4.3 Simmer Operation

Simmer operation of flash lamp is achieved by means of additional power supply of about 800V connected to flash lamp through resistance R and about 25 mA of current

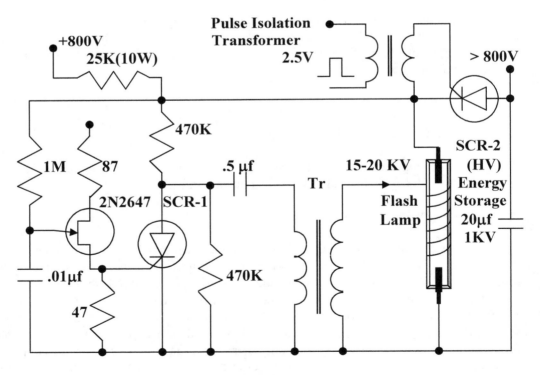

Figure-9: Simmer Operation of Flash Lamp.

is required to maintain simmer glow in flash lamp. Unijunction transistor is used in voltage sensing circuit across flash lamp to shunt trigger the flash lamp by means of atrigger transformer in whose primary, a condenser is discharged through SCR-1 as shown in figure 9. Once the simmer is produced, voltage across the flash lamp falls and no more triggering takes place, as unijunction transistor does not produce pulse at low base voltage to fire SCR-1. The energy storage condenser is discharged through high current SCR-2, which is triggered by a pulse through a high voltage isolation transformer. This type of operation is used in high repetition lasers, as it gives better efficiency and more life to flash lamp.

7.4.4 Pseudo-Simmer Operation

In pseudo-simmer operation no additional power supply is required. Simmer glow in lamp is maintained for only 50 μ seconds through a resistance connected to energy storage condenser as shown in figure 10. Flash lamp is triggered by shunt trigger circuit, to produce simmer by current supplied from energy storage condenser through resistance R. The value of this resistance R is such that about 30 mA current flows through lamp for about 50 μ second when a high current SCR-2 connected across resistor is fired, allowing the energy storage condenser to be discharge though the lamp. Operation efficiency with this operation remains same as with simmer operation. The circuit is noisy as for each operation lamp is triggered. Suppression of shunt trigger noise is required in case of Electro-optic Q-switching.

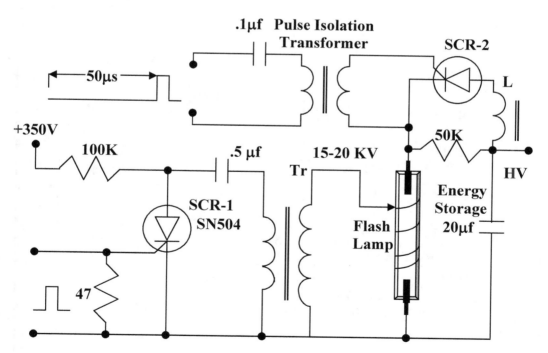

Figure-10: Pseudo-Simmer Operation of Flash Lamp

7.5 Q-SWITCHING CIRCUITS:

Synchronizing and timing circuits are used for mechanical and Electro-optic Q-switching. In rotation prism Q-switching, flashing of lamp is synchronized with prism position, so that cavity is formed when all energy from flash lamp is absorbed in the laser material and optimum population is achieved. While in case of acousto-optic, frustrated total internal reflection or Electro-optic Q-switching, r-f voltage is switched off, voltage is applied to ceramic piezo-electric spacer between prisms or high voltage in switched off /on to Electro-optic crystal at the instant of optimum inversion respectively.

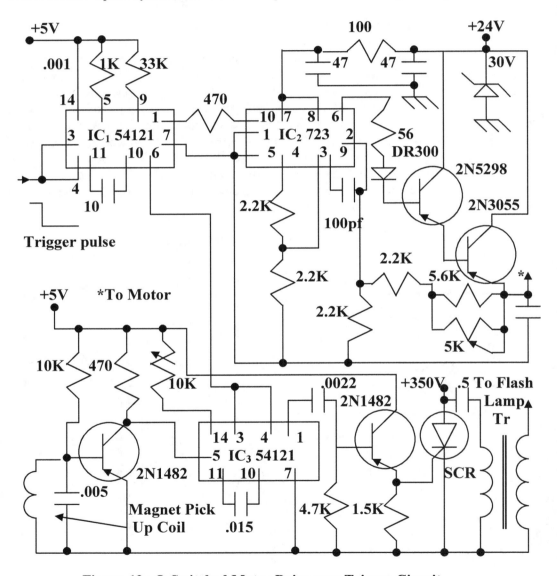

Figure-12a Q-Switched Motor Drive cum Trigger Circuit

7.5.1 Motor Drive, Magnetic Pick-Up and Delay Circuit for Rotating Prism Q-Switching

Here small voltage pulse induced on small coil as a result small magnet embedded on rotating prism mount is amplified as shown in figure-12a. When the energy storage condenser is charged with ready lamp indication, on pressing laser fire switch, a bounce free negative (-ve) TTL pulse is generated which is applied to monostable multivibrator 54121(IC_1). This monostable generate –ve and +ve TTL pulse of 300 milli-second duration at terminal 1 and 6, -ve TTL pulse switches on voltage regulator 723(IC_2) which gives drive voltage of 6V to motor on whose shaft prism is mounted. A monostable 54121 (IC_3) that can be triggered by amplified magnetic pick-up pulse is inhibited by +ve TTL pulse, till the motor drive voltage is switched off. Purpose of this operation is that with the switching off motor drive, noise from motor commutator that

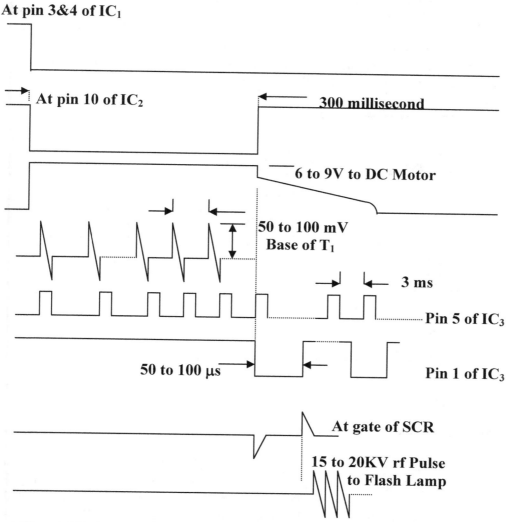

At pin 3&4 of IC_1

At pin 10 of IC_2　　　　**300 millisecond**

6 to 9V to DC Motor

50 to 100 mV
Base of T_1

3 ms

Pin 5 of IC_3

50 to 100 µs　　　　**Pin 1 of IC_3**

At gate of SCR

15 to 20KV rf Pulse
to Flash Lamp

Figure-12b:Sequence of Pulses (Q-Switched Motor Drive cum Trigger Circuit)

interferes with laser ranging is switched off and motor continue to moves with maximum speed due to inertia. At the end of this drive voltage, first amplified magnetic pick-up pulse triggers a variable delay monostable mutivibrator IC_3. At the end of this pulse, a pulse is generated to trigger SCR in lamp trigger circuit. Thus with this circuit, flash lamp can be fired at precise prism position i.e. by varying pulse width of monostable IC_3 by adjusting variable 10K potentiometer connected to pin 9 of this IC. Figure-12b shows sequence of pulses generated in this circuit.

7.5.2 High Voltage Switching Circuit for Electro-Optic Crystal

A fast and high voltage switching circuit for Electro-optic Q-switching developed at IRDE is shown in figure 13. IC monostable generates a 100 μ second or duration of flash pulse, as soon as high current SCR is fired to discharge energy storage condenser through flash lamp. At the end of this pulse a +ve TTL pulse is generated to trigger high voltage switch. The high voltage switch generates a rectangular pulse of 300 to 400 nano second duration with voltage amplitude for quarter wave phase shift voltage. This voltage applied to Electro-optic crystal generates laser energy, which remains in laser cavity. This laser energy dumped out through polarizing beam splitter at the time when quarter wave voltage on Electro-optic crystal is switched off.

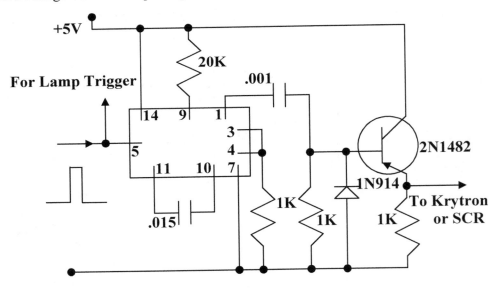

Figure-13: Delay Circuit for Electro-Optic Q-Switching

7.6 RANGE MEASURING CIRCUIT

There are two type of range measuring circuits to be used for pulse and continuous wave modulated lasers. In former type, range is measured by measuring time interval between laser transmitted and received echo pulse from target by fast counter, while in later case phase between transmitted and echo wave is compared at various modulating frequencies. In the former case range measuring accuracy of 1 meter is achieved for diffuse reflecting target, while in phase measuring method accuracy in range measurement can be achieved within millimeter for well defined target fitted with retro-reflector.

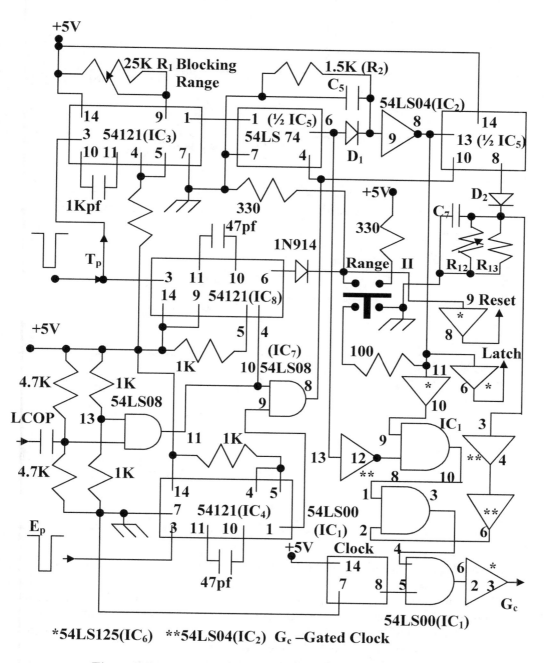

*54LS125(IC$_6$)   **54LS04(IC$_2$) G$_c$ –Gated Clock

Figure-14a Time Interval Unit (Ranging Counter)

For pulse lasers, range measuring circuit consist of electronic clock, time interval unit and fast counter to measure range with accuracy of 10, 5, 1 meter with use of clock frequency of 15, 30 and 150 MHz. The range measuring circuits developed are shown in figure below

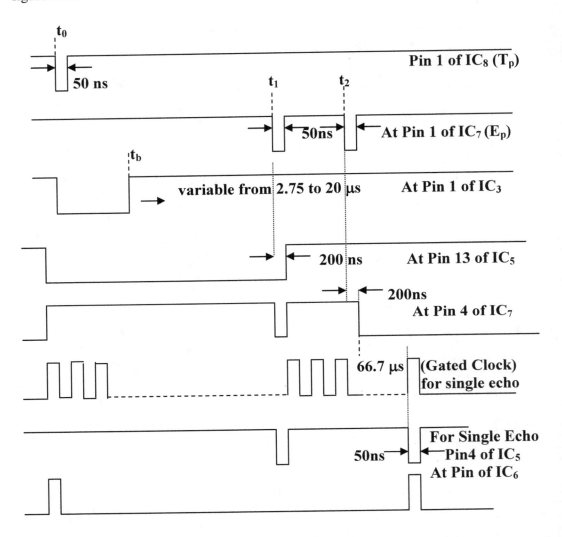

Figure-14b Sequence of Pulses at Time Interval Circuit (Figure-14a)

The range measuring unit developed to measure range [7,8] with clock frequency uses 14.998 MHz and uses 54H series IC gates, flip flops and decade counters with 5x7 dot matrix display with latch memory as shown in figure - 14a. It can display ranges of two targets in the same line of sight intercepted by laser beam i.e. 1st range is displayed while 2nd range remains in counter memory and can be displayed by pressing latch switch as shown in figure-14a. Its minimum blocking range can be varied from 400 to 3000 meters. It consists of pulse shaping multivibrator, minimum range blocking multivibrator, gating flip-flop for 1st and 2nd laser echo pulse, pulse stretcher and buffers in time interval card for remote range display and controls. It uses a hybrid clock of 14.998 MHz manufactured by M/S Vectron of USA. The transmitted pulse (T_p) shaped to 50 nano-seconds duration by monostable IC_8 (54121) is used as reset pulse for counter. The T_p pulse also generates minimum range blocking pulse with the help of monostable IC_2 whose duration can be controlled from 2.75 to 20 μ seconds with the help of blocking potentiometer (pot.) R_1(25K). This blocking pulse is used as set pulse for 1st echo flip-flop ½ IC_5, the echo pulse shaped is received as reset pulse for this flip-flop. This 1st echo gating pulse after stretching 200 nano-second by diode D_1 (1N914) and parallel combination of C_5 (100pf) and R_2 (1.5K) and 1/6 IC_2 is applied as set pulse to second flip flop. The output pulse of second flip flop correspondence to second echo pulse is again stretched by diode D_2 and parallel combination of R_{12} (1K) and R_{13} (1K) and C_7 (120pf) to 200 nano-second stops the gated clock pulse. This allows the counter to take state corresponding to second range. The resets pulse for counter, flip-flop pulse corresponding to 1st echo and gated clock pulses corresponding to 2nd echo after passing through buffer IC_6 are used for remote display. The BCD data corresponding to 2nd echo at output of counter can be transferred to same display by pressing range II switch. The waveforms at various points of time interval unit are shown in figure-14b. The counter is three-digit decade counter connected to 5x7 dot matrix display having decoded driver and latch memory in same chip.

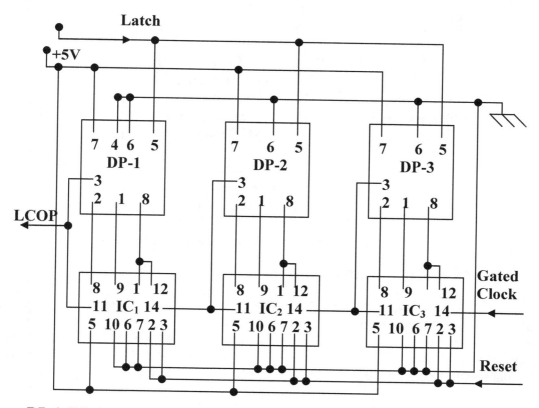

DP-1, DP-2, DP-3 hp are Type 5082-7391 and IC_1, IC_2, IC_3 are 54LS90
Figure-14c Counter cum Range Display Circuit

In recent laser range finders, ranging counter i.e. time interval unit and decade counter have been replaced by field programmable gate array (FPGA) type CPDL – XC 95288, -10C and system controller µC – AT89C51 to control laser firing with pulse coding for laser designator purpose. It uses a high-speed clock and range display for first or last echo logic as shown in figure-14d.

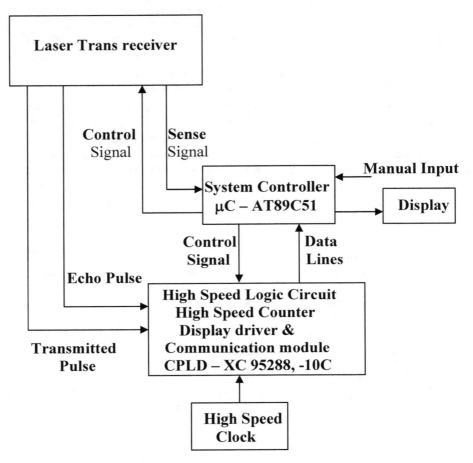

Figure –14d Block Diagram of Range Counter and Laser Control System

7.7 DISTANCE MEASUREMENT

For survey purpose, target is well defined and stationary. Therefore for accurate distance measurements, a retro-reflector can be placed on target, which also defines its exact position. For this purpose, high repetition rate low power GaAs laser is used with pulse repetition rate of 1-10 KHz or low power cw sinusoidal modulated He-Ne laser having low divergence of 50 µ radians. Three methods are used for measuring range:

Phase Measurement.
Time Measurement
Measurement by Phase Difference Elimination

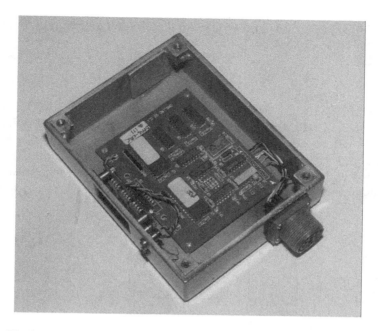

Photo-1: Ranging Counter with 150 MHz Clock (IRDE)

7.7.1 Phase Measurement

In this case, highly collimated He-Ne laser beam is intensity modulated in three frequency steps of 0.15, 1.5 and 15 MHz, and at this modulating frequencies, half wavelength correspondence to 1000, 100, 10 meters light travel distance respectively. Measurement of phase difference between modulated transmitted and reflected wave from target gives its range. The phase difference correspondence to width of pulse generated when transmitted and echo pulse crosses zero line. The width of these pulses is measured as DC voltage average over certain interval of time.

If λ_m is the length of modulated wave and n integer indicates number of half waves between laser and target, then distance D is given by relation

$$D = (n \lambda_m + \Delta\lambda_m) / 2 \qquad \dots\dots\dots\dots[4]$$

$$\Delta\lambda_m = (\phi_2 - \phi_1) \lambda_m / 360 \qquad \dots\dots\dots\dots[5]$$

Where $\phi_2 - \phi_1$ is the phase difference between transmitted and reflected wave measured by microprocessor corresponding to DC voltage.

7.7.2 Time Measurement

In this method, a low power, a high repetition rate and a narrow pulse width GaAs laser is used with pulse rate of 5 to 10 KHz. A clock of 15 MHz is used to represent one count as 10 meters. If n is the number of counts registered in the counter

172

between reference transmitted pulse and echo pulse delay, then time interval t_n as shown in figure-15 is given by

$$t_n = n\,T + t_a - t_b \qquad \qquad \text{.............[6]}$$

Where t_a and t_b are time interval between rise of range gated pulse and rise time of clock pulse, and fall of gated clock pulse and rise of next clock pulse after the gated pulse. This is determined by analyzing as voltage ratio t_a / T and t_b / T within distance measuring unit. By averaging this ratio for 1000 pulses, distance-measuring accuracy can be determined within millimeter accuracy of target range. In range measuring counter, a microprocessor monitor and control the frequency stability of the clock oscillator pulses within + or − one part per million (1ppm).

Clock Frequency – 15 MHz

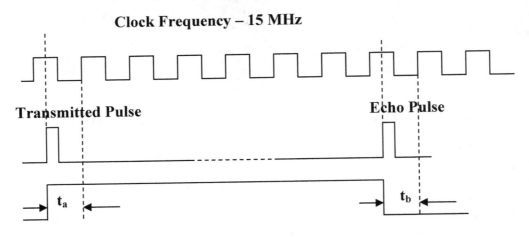

Figure-15 Range Measurement by Time Measurement Method

7.7.3 Variable Frequency Modulation

In this method, laser modulation frequency is varied, so that at consequent two frequencies, phase difference between reference transmitted pulse and echo pulse from target is zero i.e.

$$D = (\,\lambda_{n1} \cdot n_1\,) / 2 = (\,\lambda_{n2} \cdot n_2\,) / 2 \qquad \qquad \text{..........[7]}$$

$$\text{Where } \lambda_{n1} = v / f_2, \ \lambda_{n2} = v / f_2 \text{ and } n_2 = n_1 + 1 \qquad \qquad \text{..........[8]}$$

In this equation, D and n_1 are unknown, v, f_1 and f_2 can be determined D can be measured at several different zero positions through the modulation frequency range and can produce a mean value. By this method, using modulation frequency of 500MHz i.e. $\lambda_n = 0.30$ meter, a range resolution of 0.1 ppm or 0.01mm is achieved in distance. A temperature stable quartz oscillator is used. Modulation frequency is varied in step of 150 Hz. Phase of reflected beam is compared with the help of polarizing beam splitter for zero output on detector i.e. transmitted and reflected beam is in same phase and frequency is changed again till next zero is achieved between transmitted and received beam. A distance of 8 km can be measured within an accuracy of 0.5 mm by this method.

7.8 LOW LIGHT LEVEL DETECTION CIRCUITS

Low light level detection circuits are being used in laser receivers, where a very weak echo is received on photo-detectors. Photo-detectors, generally used in laser receivers are photo multiplier tubes (PMT), p-i-n photodiodes or avalanche photodiodes (APD). Following electronics circuits are used in laser receivers for lowlevel detection.

Power Supply for PMT
Bias Supply for APD
Pre-amplifier.
Amplifier

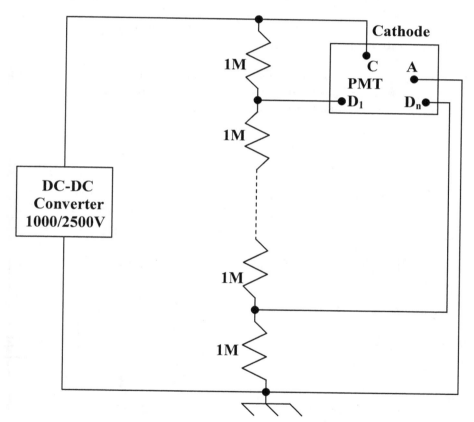

Figure-16 Power Supply Circuit for Photo Multiplier Tube (PMT)

7.8.1 Power Supply for Photo Multiplier Tube

Photo multiplier tube (PMT) are generally used for detection of laser echo when the sources in range finders operates in visible region i.e. He-Ne, ruby, second harmonics Nd:YAG. PMT needs high voltage and very low current DC power supply, generally with voltage rating from 1 to 2.5KV depending upon number of dynodes in PMT. The current requirement is less than 50 µA. Block diagram showing DC-DC inverter with voltage dividing network for biasing PMT is shown in figure-16.

7.8.2 Bias Voltage for Avalanche Photo Diode

Laser range finders operating near infrared using laser sources like Nd:Glass, Nd:YAG and Er:Glass uses silicon, germanium or InGaAs avalanche photodiodes (APD), due to high quantum efficiency at laser wavelength and internal carrier multiplication. These detectors are biased at a voltage, just below the avalanche breakdown voltage for optimum performance i.e. for carrier multiplication. But in the field condition temperature varies from –30 to +55^0 C. The avalanche break down voltage vary to a great extent as its temperature coefficient is about 2 V /^{0}C. Therefore for constant gain, bias voltage to Avalanche photo diode (APD) to vary with temperature, this can be achieved with noise or temperature control.

7.8.2.1 Noise Controlled Bias Circuit

The bias circuit is shown in figure-17, is developed for silicon APD type EMI 30500. When the detector is operated near avalanche breakdown voltage, noise at its output increases due to background and dark current. The noise is in form of pulses,

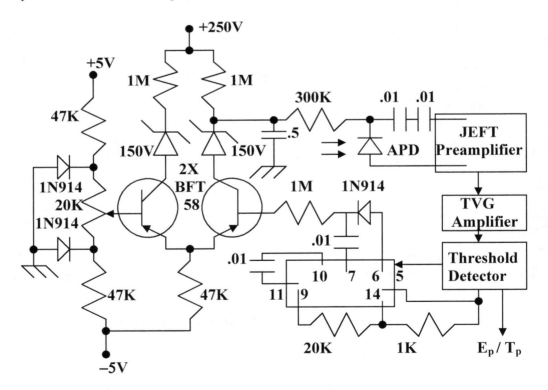

Figure-17 Noise Controlled Bias for APD

whose frequency increases with applied voltage? This noise is known as false alarm noise, as it reduces the probability of correct ranging of target. Therefore to keep this noise at fixed value as a result of background variations, dark current variation due to temperature or its gain variation as a result of its avalanche bread down voltage variation due to temperature. These noise pulses trigger a monostable multivibrator IC74121whose output after rectification is applied to base of transistor T_1 of differential amplifier, whose collector is connected to high voltage through resistor R and collector voltage of transistor T_1 bias silicon APD. This voltage decrease with APD noise, reducing gain, hence keeping the noise at constant value due to variation in background noise, dark current / break down voltage with temperature. This bias circuit gives constant false alarm probability of correct ranging as a result of background or temperature variation.

7.8.2.2 Temperature Compensated Time Variable Bias Circuit

This circuit as shown in figure-18 is developed at IRDE to bias reach through structure silicon APD, RCA type C30954E, where dark current noise or micro plasma

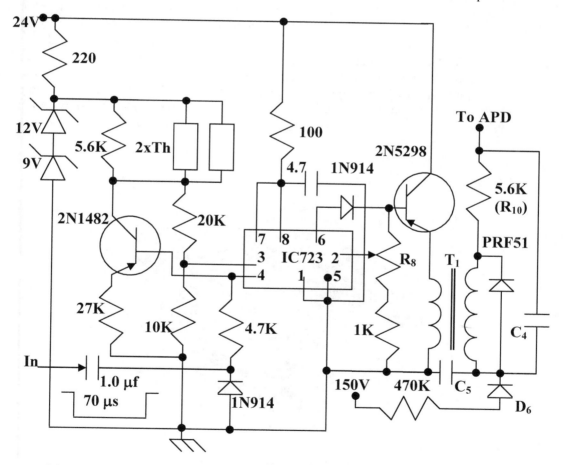

Figure-18 Temperature Compensated Time Variable Bias for APD

noise is absent till detector reaches to break down voltage. The detection capability is limited only due to background noise. The bias voltage increases with time, at the instant of laser transmission till almost breakdown voltage, corresponding to time for receipt of laser echo from maximum range i.e. 66.67 μ second for range finder with maximum range capability of 10 km. Thus due to variation in bias, gain of APD varies with time, giving constant output for laser echo received from near or far targets. This bias circuit not only improves correct detection probability, but APD damage due to strong backscatter from aerosol or strong reflection from near target is avoided. It consists of constant current source supplying current to a network of thermistors and resistances such that voltage develops across this network varies in constant ratio to that of avalanche break down voltage of APD with temperature i.e. at point P shown in circuit diagram. This voltage is applied to non-inverting terminal of voltage regulator IC 723, which is operated in, shut down mode. At the instant of laser transmission, 70 μ second negative TTL pulse generated is applied to current limiting terminal of IC 723, so that output is available at output terminal whose value can be adjusted by potentiometer R_8 (5K). This output is applied to base of emitter follower with its emitter connected to primary of pulse transformer T_1. This voltage is stepped up 20 times and is applied to charge condenser C_4 (1Kpf) through resistance R_{10} exponentially. A DC off-set voltage of about 100 to 150V, depending upon break down voltage of silicon APD is applied at the condenser C_5 (1.0 μf) connected in series with condenser C_4 (1Kpf) through diode D_6 (PRF51). This DC voltage with exponentially variable voltage applied to APD gives time variable gain to APD whose output is connected to pre-amplifier. The output voltage can be adjusted by R_8 so that its maximum value should be little less than break down voltage of APD

Photo-2: Temperature Compensated Time Variable Bias (IRDE)

7.8.3 Low Noise Pre - Amplifier

APD have high input impedance, therefore for proper impedance matching, a high input impedance amplifier is required with low output impedance. A trans-impedance amplifier with input impedance of 4.7/10KΩ with bandwidth of 40 MHz with output impedance of 50Ω as shown in figure-19. The amplifier uses two UHF transistors 2N918 with gain bandwidth product of 1GHz. The pre amplifier is low noise giving background limited noise limitation to range finder with 1 milli-radian directivity with receiver aperture of 40 mm. The performance of preamplifier depends on circuit lay out.

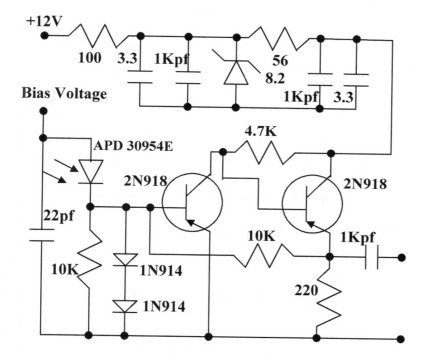

Figure-19 Trans-Impedance / Preamplifier for APD

7.8.4 Amplifier

The output of pre-amplifier is coupled to high gain amplifier developed at IRDE as shown in figure-20. It has a voltage gain of 400 with bandwidth of 40 MHz. This circuit also uses two UHF transistor 2N918 connected in cascade, condenser coupled with resistive load only. Its lay out in also very important to avoid feedback in order to avoid instability / oscillations.

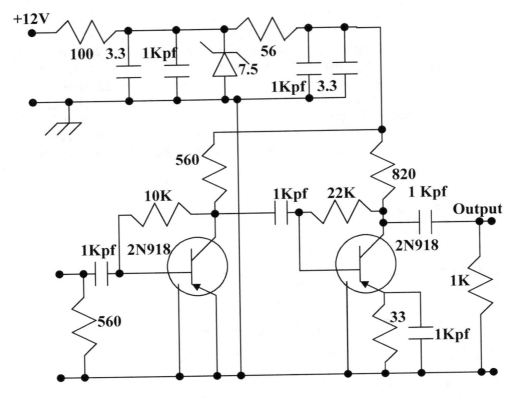

Figure-20 Pulse Amplifier Circuit

7.9 SAFETY CIRCUITS

The safety circuits are very important in laser range finder for personnel and component safety. The following safety devices / circuits are used.

High voltage energy storage condenser is always kept short circuited through normally closed contacts of high voltage relay for personnel safety.

In case of failure of high voltage sensing circuit in fly back inverter, a clamping transistor with zener diodes to clamp high fly back voltage for charging energy storage condenser beyond its maximum rating for safety of personnel and components. Over voltage protection at 5.2V in +5V switching regulator using IC voltage comparator with fuse in input circuit gives protection to TTL IC circuits rated for maximum 5.5V.

The silicon APD is protected from intense back ground or laser scatter from atmospheric aerosol, in case of large receiver aperture or high repetition rate lasers. Using low bias at APD at time of laser transmission of laser pulse from range finder, this voltage increases with time to a voltage just a little value less than avalanche break down voltage

SUMMARY

Electronics design plays a very important role in laser range finders, because the instrument reliability depends on component, layout of circuits, wiring and efficiency of power supplies and inverter for charging energy storage condenser. Safety circuits are required for component and personnel safety. Flash lamp life, its efficiency depends upon trigger methods used and its impedance matching with pulse discharge circuits. Laser pulse reproduction depends on synchronizing circuit. This chapter gives all aspect of circuit design carried at IRDE for various circuits like fly back inverter, trigger circuits, low voltage regulated supplies, range counters and low-level detection circuits. M/S Bharat Electronics Ltd. Bangalore has translated some of the circuits using discrete components into thick film hybrid circuits. These are BMC-1273 Range processor, BMC-1274 Range Counter, BMC-1277 Receiver Amplifier and BMC-1275 Timing and bias Pulse Generator.

REFERENCES

1. Don Latshu, "Optimal Design of Converter circuit for Portable Laser Range Finder", Thesis submitted for partial fulfillment of Master of Science in Engineering. Philadelphia, Pennsylvania, USA, (Dec. 1969).

2. Juyal, D.P., "A Compact high Power Pulsed TEA CO_2 Laser", J. of Telecom. Engineers. (India), Vol. 33, No. 6, June 1987, pp. 186.

3. Gandhi, S.K., Mansharamani, N and Rampal, V.V., "Double Pulse driver for GaAs Laser Diode", J. of the Instruments society of India, vol. 7, No. 4, December, 1977, pp. 1-2.

4. M/S ILC technical Bulletin, "An Overview of Flash Lamps and Arc Lamps", California, USA.

5. Juyal, D.P., "Design of an R.F. Excited helium-Neon Visible Gas Laser and Study of the Optimal Condition for Gas Mixtures and Pressures", Defence Science Journal (India), Vol.22, No. 5, October 1972, pp.245-248.

6. Mansharamani, N., "Q-Switching Techniques for Portable Laser Range Finder", Presented at symposium on Infrared Technology and Instrumentation, held at Bhaba Atomic Research Center, Bombay, (April, 1980).

7. Mansharamani, N. and Rampal, V.V., "A Compact Time Measuring Unit for Use with a Laser range finder", J. of Physics and Scientific Instruments, Vol. 7, May 1974, pp. 966-968.

8. Mansharamani, N. and Rampal, V.V., "Time Measuring in Laser Ranging systems", Proc. Of Seminar on Time and Frequency, pp. 320-323, Publisher INSDOC, New Delhi-110012, (1977).

9. Price, W.F. and Uren, J., Laser Surveying", Publisher-Van Nostrand Reinhold (International) Co. Ltd., London 1989).

10. Mansharamani, N., "Noise Controlled Bias System for Silicon Avalanche Photodiode", J. of the Institution of Electronics and Telecommunication Engineers (India), Vol. 25, No. 6, (June 1979).

11. Edwards,B.N., "Optimization of Preamplifier for Detection of Short Pulses of Light with Photodiodes", Appl. Optics (USA), Vol. 5, No. 9, September 1966, pp. 1423-1425.

12. Bobraty,B.E. and Farnsworth,R.P., "Constant False Alarm Rate Bias Control for an Avalanche Photodiode Laser Receiver", The Review of Scientific Instruments, Vol. 41, No. 8, August 1970, pp. 1191-1195.

CHAPTER-8

OPTICS & THIN FILMS

8.1 INTRODUCTION

Optical design, thin film and optical technology play very important role for good performance of laser range finders. For efficient laser source, the requirement of optical components is

1. Optical glass material from which optical component are fabricated should be highly homogeneous, free from sleeks and bubbles, homogeneity should be interferometeric grade i.e. equal or better than ¼ fringe shift per inch travel of light in glass.

2. Glass used should have high transmission at laser wavelength, better than 99.5% per cm. travel of light.

3. Material should be hard to give surface finish without any digs, to a surface figure of $\lambda/10$ or better.

4. It should have good heat conduction capacity and should be able to withstand laser power of at least 100 MW/cm^2.

5. Glasses should be non-hygroscope and atmospheric gases or atmospheric impurities should not affect their surface figure.

6. Glass quality should not be affected by surface cleaning procedure followed before applying multi-layer thin film coating- reflecting or anti reflecting.

7. Material / glasses used should give good adhesive to multi-layer coating.

The glasses generally used for laser operating in visible are fused silica, borosilicate glass / BK-7 or sapphire. For CO_2 laser wavelength, germanium or zinc selenide (ZnSe) / zinc sulfide. Similarly glasses used for laser receiver or for common optics for receiver in receiver cum sight should have good transmission for laser wavelength.

Optical components used in laser range finder are, prisms, dielectric mirrors, resonant reflectors, windows, wedges, lenses, colored filter glasses, polarizing beam splitters, quarter wave plate, Electro-optic crystal, beam splitters for separating laser and visible wave length. All optical components used in giant pulse laser source, including first lens of beam expanding telescope / laser collimator for ranging non-cooperative target should be capable of withstanding high peak power of the order of at least 50 MW / cm.^{2}

In this chapter, design philosophy followed at IRDE for optical computation for compact design of laser collimating optics, large area receiving optics including sighting / aiming optics is given and few design are illustrated in figure-1 and -2.

Specification of glass, laser component and materials for Q-switching with fabrication and testing procedure followed are briefly described along with references of books and publication for detail study in this field if required. The optical components fabricate at IRDE and optical testing facilities are illustrated in photographs. Specifications of trade items purchased are also given.

In Thin-Film Laboratory of IRDE, material used, cleaning and coating procedure followed including testing for laser mirrors for reflectivity and power handling capacity, coating procedure for beam splitter used in sighting cum laser receiver are also described in brief. Materials and procedure followed for high damage resistant anti reflection coatings on laser components, prisms, aligning wedges, lenses for transmitting and receiving optics is also given.

Specification and testing of trade items, like laser rods, Q-switched component for passive and Electro-optic methods, filters both colored and interference, resonant reflectors along with manufacturer address is also given.

8.2 OPTICAL COMPUTATION

Design Aspects of Laser Based Optical Systems

In laser instrumentation laser collimator and laser receiver or laser receiver cum sight are main optical systems, which are commonly required. Laser collimators are basically Galilean telescope, which is used to reduce the beam divergence of raw laser beam by its magnification times. Usually in the design of Galilean telescope both positive and negative element are corrected but this increases the over all length of the system. As the spectrum and field of view covered is small, positive element can be used at very low focal number (f/no.) and its aberration can be balanced by the negative element. This helps in reducing the overall length. Care should be taken for the radius of the last surface of the negative element so that its antireflective light does not focus on front surface of the laser road. The design data of compact laser collimator of 4X for laser beam of 5mm diameter is shown in figure-1.

Laser receivers are another common optical system used in laser range finders. It has following typical requirements:

* Interference filter is to be used which should be placed in approximately parallel light.

* Diameter of the detector active area used is normally 0.8 mm, which is covered by a thin glass plate of the size 0.3mm, which should be considered in design for low f number receiver, as it contribute to 3rd order aberrations.

*Field of view is another very important parameters, which decides the design approach and the type of receiver to be used. Field of view is decided by the target to seen by the receiver. Typically it is of the order of 2 mrad or less. This makes its focal length 400 mm or more.

4X Laser Beam Expander

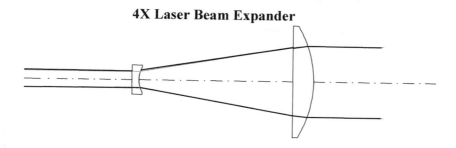

	RDY	THI	GLA	Aperture Dia.
OBJ:	INFINITY	INFINITY		
STO:	151.94019	2.00	BK7_SCHOTT	5.00
2:	7.66153	44.09		
3:	INFINITY	6.00	BK7_SCHOTT	20.00
4:	-32.42463	5.00		
IMG:	INFINITY	INFINITY		

SPECIFICATION DATA

Entrance pupil diameter	5.00 mm
Wavelength	1064.00 nm
Angle	2 milliradian
System Length	52.00 mm

Figure-1: Laser Collimator with Magnification X4

Receiver aperture is decided by the range requirement of the instrument. The laser receiver design data for laser receiver with 50 mm aperture and directivity of 2 milliradians are shown in figure-2. Depending upon these parameters following configuration can be considered. A Galilean telescope followed by interference filter and then a focusing element. System then can be optimized to get the focus on the glass plate of the detector, which is around 1.6 mm before the detector. A pinhole can be placed on this glass plate to control the FOV precisely. If field of view of the order of 1 m radians. or less, a Cassegarrian type of catadioptric system should be used. If these receivers are be combined with optical sight then a Porro prism combination with dichotic coating or roof Penta combination is to be used.

LASER RECEIVER

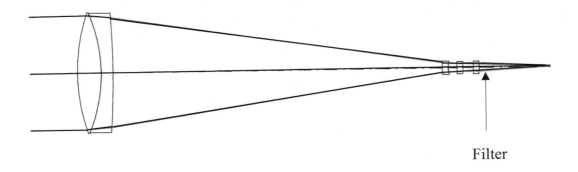

Filter

	RDY	THI	GLA	Aperture Dia
OBJ:	INFINITY	INFINITY		
STO:	82.83067	10.000	510644_CHANCE	50.0
2:	-70.11248	5.0000	620364_SCHOTT	50.0
3:	-370.21890	139.320766		
4:	INFINITY	2.486859	510644_CHANCE	5.0
5:	6.07220	3.733492		
6:	INFINITY	2.490318	BK7_SCHOTT	5.0
7:	INFINITY	4.481705		
8:	INFINITY	2.486859	SK16_SCHOTT	5.0
9:	-18.23963	30.000000		
IMG:	INFINITY	0.000000		

SPECIFICATION DATA

Entrance Pupil Diameter	50 mm
Wavelength	1064 nm
Detector Size	**0.8 mm**
Field of View	2 mrad
EFL	400 mm
System Length	200 mm

Figure-2: Optical Design Data of Laser Receiver with 50 mm Receiver Aperture.

8.3 OPTICS TECHNOLOGY

Optics technology used for fabrication and testing of optical component is different in sense that fabrication techniques used in fabrication of component used in laser cavity requires high grade optical homogeneous glass i.e. interferometeric grade / Schlieren grade fused silica or BK-7 silicate glass. Surface finish should be better than $\lambda/10$ or better with cosmetic surface quality, according to U.S. Military Surface Quality Specification, MIL-0-13830A. The scratch dig designation for laser's optical component is 10-5 i.e. dig diameter 0.1mm or less and sum of maximum visible scratches are present, their combined lengths cannot exceed a quarter of the part diameter. For clear aperture 0-20 mm, which is 90% of total optical component diameter. Component surface should have maximum one dig and the maximum sum of dig diameters is 2 times the maximum dig diameter of 0.1 mm. For clear aperture diameter of 20-40, 2 maximum are allowed, and the maximum sum of dig diameter is 4 times the maximum dig diameter of 0.1 mm.

The optical components used in laser are TIR prism, corner cube prism, wedges, plane and large radius focal length concave, partial and total reflecting mirrors. While other component used are, narrow band interference filter, resonant reflector, Nd:YAG laser rods, $LiNbO_3$ Q- switch crystal, polarizers, quarter wave plate purchased as trade items. The specifications of trade items used as well as the specification of other optical component like lenses, prism, graticules used in collimating, receiving and sighting telescope fabricated at IRDE are given.

8.3.1 Fabrication of Laser Rod

Laser material i.e. Nd:Glass in the form of rectangular blanks of Central Glass & Research Institute, Calcutta (CGCR&I) is fabricated in the form of laser rods with fine grind cylindrical surface in required diameter, using center less grinder. These laser rods with roughly cut ends are held in an aluminum cylindrical adopter having holes almost equal to the diameter of the rod. The length of the metallic adopter is smaller than the laser rod length, so those support pieces are fixed around both ends. Having properly fixed the rod with a suitable wax in the adopter, both the ends of the rod are polished, usually by hand touching to get the required surface accuracy and cosmetic figures on the end faces. Parallelism of the end faces is also simultaneously maintained during polishing. Surface accuracy is tested with a laser interferometer and parallelism of the faces could be tested with autocollimator / interferometer. Specifications of optical finish achieved at are

Size:	6.25mm diameter, 80mm long
Surface Finish:	$\lambda / 10$
End Configuration:	Plane-parallel within 2 seconds.
Perpendicularity:	within 1 minute

8.3.2   Fabrication of TIR Prism

Prism blanks from interferometeric grade fused silica blank in form of right angle prism i.e. Spectrosil A grade fused silica from M/S Thermal Syndicate of UK in a size slightly larger is taken. TIR prism is fabricated in a similar manner as done for normal prisms. For achieving high order of surface accuracy, runners are used on the faces to hold the prism during polishing. Angles of TIR prism are controlled by testing the through angle and pyramidal angle by internal reflection techniques using an Angle Dekkor. The specifications of TIR prism achieved are

Size of Prism:	10x10 mm – entrance face
Clear Aperture:	8x8 mm
Surface Figure:	$\lambda/10$
Surface Quality:	10 – 5, MIL-O-13830A
Angle Accuracy:	90^0 +/- 2 seconds of arc.
Pyramidal Error:	< 2 second
Bevel on 90^0 edge:	0.02 mm or less.

8.3.3   Fabrication of Corner Cube Prism

The corner cube prisms fabricated from cube blanks of high quality fused silica of infrasil grade procured from M/S Heraeus of West Germany. These prism are required as total reflector in compact Nd:YAG passive Q-switched laser source for hand held range finder. The advantage in using corner cube prism is that defective passive Q-switched element can be replaced in laser cavity without disturbing alignment of laser Trans-receiver unit. The requirements are very stringent on angle accuracy i.e. within 2 seconds, so that deviation of incident beam and reflected beam is within 5 seconds. The size of cube blank taken is 2.8 times the size of prism aperture plus cutting allowance. The method used for preliminary fabrication and polishing is similar as given in reference [2] i.e. corner of a cube to be polished to a high degree of angular and surface accuracy before slicing four prisms from a cube. The specifications of corner cube achieved are

Material:	Fused silica interferometric grade.
Size:	10mm.
Clear Aperture:	8.5mm.
Angle Accuracy:	within 2 seconds
Surface Finish:	$\lambda/10$
Surface Quality:	10 – 5, MIL-0-13830A.

8.3.4   Fabrication of Flat and Concave Mirror Substrate

Mirror substrates are generally used are from high optical quality fused silicate or from boron silicate glass BK-7 that is hard, non-hygroscopic gives very good surface finish. This glass has very less absorption coefficient for visible and near infrared light i.e. requirement for partial mirror. For CO_2 laser germanium substrate is used for partial mirror, while metallic mirror with gold coatings are used as total reflector in a laser cavity. The minimum diameter of substrate is 1.2 times the clear aperture of required and minimum thickness is 1/6 of diameter. Large radius curved mirrors are for He-Ne lasers and for unstable resonator radius of curvature depends on cavity length. The surface of mirror should be polished to a figure of $\lambda/10$ or better. Surface quality: 10 – 5 scratches and digs i.e. MIL-0-13830A.

8.3.5 Fabrication of Wedge Plates

The fabrication of wedge plate is similar to the fabrication of flat surface. The only difference is of the generation of a wedge plate, critical angle on the surface of the plate could be generated during grinding and smoothing and final wedge angle could be controlled during polishing using an Angle-Dekkor. Detail fabrication is given in reference [2].

8.4 SPECIFICATIONS OF CRITICAL COMPONENTS:

The specifications of some critical components like interference filter, lithium niobate, quarter wave plates; polarizing beam splitter and Q-Switched acetate sheet used in laser range finders are given below

8.4.1 Interference Filter: Interference filter is used in laser receiver to allow only laser echo to be received on detector.

Size: 12.5 mm diameter
Aperture: 9.5mm
Surface figure: $\lambda/4$ or better
Transmission: 40% minimum at laser wavelength.
Band width: 100 A^0
Blocking from x- ray to far infrared, with integrated transmission in blocking region less than 0.1% of total integrated transmission in pass band.

8.4.2 Air Spaced Polarizing Beam Splitter Cube: Used in Electro-Optic Q-Switching.

Material: BK-7 Grade A
Surface Quality: 20-10.
p-Polarization Transmittance (T_p) > 98%.
s-Polarization Reflectance (R_s) > 99%.
Contrast Ratio (T_p / R_s) > 500:1
Damage Threshold is > 500 MW / cm^2 for 20 ns pulse at 1.064 μm.
Acceptance Angle: $0^0 - 1^0$.
Transmittance wave front distortion < $\lambda/4$.
Transmitted beam deviation < 2 minutes.
Reflected beam Deviation < 4 minutes.
Clear Aperture: 8 mm.
Operating Temperature: -40^0 C to 70^0 C.
Humidity: 95%.

8.4.3 Lithium Niobate

Size: 9 x 9 mm and 25 mm long.
Surface Figure: $\lambda / 10$ or better.
Coatings: High field damage resistant anti reflection coatings.
Damage Threshold > 50 MW / cm^2 for 20 ns pulse at 1.064
Transmittance wave front distortion < $\lambda / 4$.
Clear Aperture: 8 mm.
Operating Temperature: -30^0 to 55^0 C
Humidity: 95%.

8.4.4 Quarter Wave Plate:

Material: Quartz Single Crystal.
Diameter: 12.5 mm
Thickness: 6mm.
Clear Aperture: 9 mm.
Retardation Tolerance: $\lambda / 500$.
Wave Front Distortion: $\lambda / 10$.
Damage Threshold is > 500 MW / cm^2 for pulse of 5 ns.
Temperature Coefficient: 0.0001 $\lambda / {}^0$C.

8.4.5 Eastman Q-Switch Acetate Sheet (Cat. No. 15064)

Material: Bis[(4-dimethyaleminodithiobenzil(nickel)] in cellulose acetate.
Optical Density: 0.36 $\pm$ 0.02 at 1.06 micron
Nominal Size: 13 by 16.5 cm
Surface Quality: 24 fringes or better per square cm at Sodium D line
Surface Quality: Visual inspection whereby no more than two defect per sheet in less than 5% of the area are allowed. Visual imperfections will be marked
Power Density (Peak): 300 megawatt per square centimeter

8.5 THIN FILM TECHNOLOGY

The application of thin film coatings on optical components or substrate is to reduce reflection losses with help of low absorbing dielectric material of different dielectric constant depending upon refractive index μ_s of component or substrate. The coating given to reduce reflection losses is known as antireflection coating. The thin film coating given to glass or metal substrate to increase its normal reflectivity is known as reflecting coating. The antireflection coatings are given to optical component transmitting visible or laser light. While multilayer coatings of high and low refractive index dielectric material with low absorption coefficients are give given to glass substrate to reflect laser light or partially reflect and transmit laser light in laser cavity. The dielectric coated mirrors are low losses and at the same time can withstand high peak power in giant pulse lasers. The multi-layer dielectric coatings are given band pass color glass substrate to reduce bandwidth of light to be used before detector for filtering background radiation. These types of filters are known as interference filters.

If μ_s is the reflectivity of substrate and μ_1 is the refractive index of material whose $\lambda / 4$ thick layer is deposited on substrate, then reflectivity R of light at wavelength is given by relation

$$R = \left(\frac{\mu_s - \mu_1^2}{\mu_s + \mu_1^2} \right)^2 \qquad \ldots\ldots\ldots\ldots\ldots[1]$$

Depending upon the refractive index of substrate and coating material refractive index, single or two layers can reduce reflection losses from 4% to 0.1% or at the most two layers can reduces losses drastically. For two layers with refractive index μ_1 and μ_2 of optical thickness $\lambda / 4$, the reflectivity reduced is given by relation

$$R = \left(\frac{\mu_2^2 - \mu_1^2 \mu_s}{\mu_2^2 + \mu_1^2 \mu_s} \right)^2 \qquad \ldots\ldots\ldots\ldots[2]$$

For example, single layer $\lambda /4$ coating of MgF_2 with refractive index of 1.38 reduces reflection of Nd:YAG laser rod with refractive index of 1.82 to 0.1%, while for glass with refractive index normal reflectivity is reduced to 1.4%. With two layers, the refractive index of material chosen with relation

$$(\mu_2 / \mu_1)^2 = \mu_s \qquad \ldots\ldots\ldots\ldots[3]$$

Gives zero reflectivity.

For high reflectivity coating, thin film coating of $\lambda / 4$ from material with high and low refractive index material and to start and end with a high refractive index material, so that structure have an odd number of layers l, we get reflectivity as

$$R_{max.} = \left(\frac{\mu_1^{l+1} - \mu_2^{l-1} \cdot \mu_s}{\mu_1^{l+1} + \mu_2^{l-1} \cdot \mu_s} \right)^2 \qquad \dots\dots\dots[4]$$

The coating material should have good transmission at the wavelength of required light, capability to withstand high peak power and not affected by atmosphere or its pollutants. For dielectric mirror maximum reflectivity achievable is between 99.5 to 99.8% depending on coating material, number of layers and coating procedure

The procedure for thin film coating on optical component is divided into three parts

*Substrate cleaning procedure and selection of dielectric material.
*Degassing of substrate and coating process.
*Testing of coated component.

Selection of dielectric material: The coating materials used are

(i) ZrO_2 – high index and SiO_2 –low index for antireflection coatings at 1.06 µm on optical components like lens, prisms, aligning wedges etc. The transmission achieved is 99%.

(ii) TiO_2 – high index and SiO_2 – low index for partial mirrors at 1.54µm i.e. 70%, 80%, and 90% reflecting.

(iii) ThF_4 – low index and ZnSe – high index for anti reflection coatings at 10.6 µm on ZnSe substrate. Transmission achieved is 99.2%.

(iv)   Au for high / total reflecting mirrors at 10.6 µm on Cr / Cu substrate. Reflectivity achieved is 98.5%.

Before coating, the substrates are subjected to the following cleaning procedures

(a)   Ultrasonic cleaning for 10 minutes using Teepol solution in deionized water.

(b)   Washed in deionized water for ultrasonic cleaner. This step is again repeated for another 10 minutes.

(c)   Cleaned in isopropyl alcohol for 10 minutes in ultrasonic cleaner. This step is again repeated for another 10 minutes.

(d)   Cleaning by vapor degreaser using isopropyl alcohol for about 40 minutes.

8.5.1 Coating Process

Parameters

Deposition Method:	Thermal evaporation by electron beam gun (E.B.G.) system.
Substrate Temperature:	250^0 C.
Job Rotation:	30 revolution per minute (RPM)
Working Pressure:	5 to 6 mbar.
Thickness Control:	Quartz Crystal Monitor.

The cleaned substrate / optical component to be coated are loaded in vacuum chamber of the coating plant and chamber evacuated to a pressure of $2x \ 10^{-6}$ mbar with the help of rotary and diffusion pumps. The substrates are heated up to 250^0 C for about 2 hours before coating. Then just before coating the substrates are subjected to glow discharge cleaning for about 10 minutes at pressure of 10^{-2} mbar. Again when the pressure comes to $2x10^{-6}$ mbar, which is achieved faster with the help of liquid nitrogen cooling of pump system, the coating material placed in the crucibles of E.B.G. system are degassed for some time. When the final pressure of $2x10^{-6}$ mbar is settled, the coating materials are evaporated with E.B.G. The oxide materials are deposited in O_2 partial pressure of about $5x10^{-4}$ mbar. Quartz crystal monitor monitors the film thickness and rate of deposition. All parameters like film thickness, rate of deposition, glow discharge time, power to the source etc is fed to the microprocessor, which controls the complete process. During the complete coating cycle the job (substrate) is roared at a speed of 30 RPM in order to get uniform coating. After the substrates are coated, they are allowed to cool down to room temperature and then taken out of coating chamber.

SUMMARY

The performance of laser range finder depends upon optical design, optical and laser material and component, fabrication techniques, thin film materials and coating processes. As these components are critical components, cost of a laser range finder depends on cost of these components. In this chapter, starting with optical materials, fabrication techniques, testing methods of TIR prism, corner cube prisms, and optical flats for mirrors, windows and wedges followed at IRDE are described. Fabrication method of laser rods from Nd:Glass material developed at C.G.&C.R.I is also given. The cleaning methods of optical and laser components are utmost important, as life of instrument depends of cleanliness of these component, therefore assembly of laser is carried out in a dust free clean room. The thin film coating procedures followed at IRDE are also described in this chapter.

APPENDIX

POLARIZATION

Light beam is said to be linearly polarized, if its direction of electric / magnetic field remain fixed in a particular direction. For unpolarized wave, direction of field fluctuates in a random manner. Partially polarized light is mixture of polarized and unpolarized light.

Circular and Elliptical Polarization

A circularly polarized wave is interpreted as two linearly polarized waves of same amplitude E_0 polarized orthogonally to each other or a single wave in which the electric vector at a given point is constant in magnitude but rotates with angular frequency ω. A wave is right or left circularly polarized, if rotation of electric vector is clock or counter clockwise when viewed against the direction of propagation.

Elliptical Polarization: If the component (real) fields are not of the same amplitude, the resultant electric vector, at a given point in space, rotates and also changes in magnitude in such a manner that the end of the vector describes an ellipse. Circularly polarized light is produced by plate fabricated from doubly refracting transparent crystals such as calcite or mica.

Relation gives thickness d of a quarter-wavelength plate

$$d = \frac{\lambda_0}{4(n_1 - n_2)}$$

Where n_1 and n_2 are refractive index of axis for slow and fast orthogonal wave with slow and fast velocity of double refracting crystal. Circularly polarized light can be produced using electro-optic crystals.

Quarter-wave plate: An optical element that converts polarized light into circularly polarized light.

Half-wave plate: An optical element that converts linearly polarized light into orthogonally oriented linearly polarized light.

REFLECTION and REFRACTION

1. Laws of Reflection

Angle of Incidence is equal to angle of refraction

2. Law of Refraction (Snell's Law)

If θ is angle of incidence in a medium with refractive index n_1 and angle of refraction ϕ in medium with reflective index n_2 then

$$n_1 \sin\theta = n_2 \sin\phi \text{ or } \sin\theta = n \sin\phi$$

where $n = n_1 / n_2$

For polarized ray with electric vector in plane of reflecting/refracting medium or perpendicular to plane of incidence, ratio R_s of amplitude of electric vector of reflected to incident wave is given by

$$R_s = \frac{\cos\theta - n\cos\phi}{\cos\theta + n\cos\phi}$$

For polarized ray with electric vector in plane of reflecting medium, ratio R_p, amplitude of reflected to incident wave is given by

$$R_p = \frac{-n\cos\theta + \cos\phi}{\cos\theta + \cos\phi}$$

For normal incidence, $R_s = R_p = (1 - n) / (1+n)$

Frensel's Equation: By the use Snell's refracting law

$$n = \sin\theta / \sin\phi$$

Amplitudes of refracted T as well as reflected ray R normalized with respect to amplitude of incident wave may be expressed as

$$T_s = \frac{2\sin\phi\cos\theta}{\sin(\theta + \phi)}$$

$$R_s = -\frac{\sin(\theta - \phi)}{\sin(\theta + \phi)}$$

$$T_p = \frac{2\sin\phi\cos\theta}{\sin(\theta + \phi)\cos(\theta-\phi)}$$

$$R_p = - \frac{\tan (\theta - \phi)}{\tan (\theta + \phi)}$$

Amplitude of reflected ray in term of refractive index is expressed as

$$R_s = \frac{\cos \theta - (n^2 - \sin^2 \theta)^{1/2}}{\cos \theta + (n^2 - \sin^2 \theta)^{1/2}}$$

$$R_p = \frac{- n^2 \cos \theta - (n^2 - \sin^2 \theta)^{1/2}}{n^2 \cos \theta + (n^2 - \sin^2 \theta)^{1/2}}$$

For normal incidence, intensity of reflected beam for both direction of polarization is given by

$$I_{ref} = (n - 1)^2 / (n + 1)^2$$

External and internal reflection

For external reflection n > 1

For internal reflection n < 1, in this case, at certain angle, known as critical angle for which refractive angle is 90^0 is known as critical angle

$$\theta_{critical} = \sin^{-1} n$$

Polarizing angle: For polarized light with polarization in plane of reflection and refraction, refection is zero at certain angle known as Brewster angle and is given by

$$\theta_{critical} = \tan^{-1} n$$

No Reflection

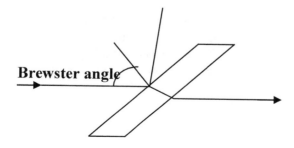

Figure-1A Brewster Window

Glass plates placed at Brewster angle to light beam are known as Brewster windows. Brewster windows are used in laser cavity for getting polarized light with no reflection loss.

COHERENCE and INTERFERENCE

If Δv is bandwidth of a light with frequency v and wavelength λ, then coherent length L_c is given by relation

$$L_c = c / \Delta v = \lambda^2 / \Delta \lambda$$

Where $\Delta \lambda$ is the width of the spectral line

For example for laser with frequency of 10^{14} bandwidth 10^3 Hz coherent length is 10^{11} wavelengths or about 50 km

Optical interference of light was first demonstrated by Thomas Young in 1802.

The theory of optical Interference is based on principle superposition of electromagnetic fields E produced at a point in space.

$$E = E_1 + E_2 + \ldots\ldots.$$

Where $E_1, E_2, \ldots$ Are fields produced at a point, Intensity at a point is given by

$$I = I_1 + I_2 + 2 E_1 . E_2 \cos \theta$$

The Michelson interferometer developed in 1880, described the versatile interferometer device used for measurement of refractive index of gases. A modification of the Michelson interferometer known as the Twyman-Green interferometer is used for testing optical elements such as lenses, mirrors and prism. Collimated light is used in this case and optical elements are placed in the path as shown in figure-2A. Imperfections are visible by distortion in the fringe pattern.

Interference with Multiple Beams

The most common method of producing a large number of mutually coherent beams is by division of amplitude. The division occurs by multiple reflections between two parallel, partially reflecting surfaces. These surfaces might be semitransparent mirrors, or merely the two sides of a film or slab of transparent material

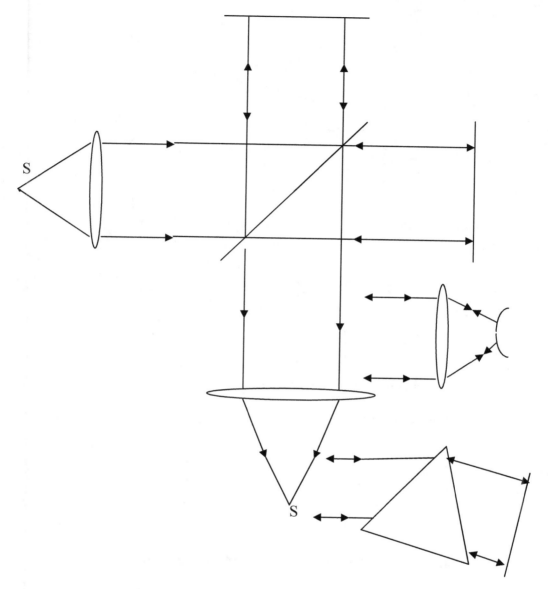

Figure-2A: The Twyman-Green modification of the Michelson interferometer.

The Fabry-Perot Interferometer

Interferometer devised by C. Fabry and A. Perot in 1899 is used to measure wavelength with high precision and to study fine structure of spectrum lines. It consist of two optically flat partially reflecting plates of glass or fused quartz with their reflecting surfaces held accurately parallel. The plate space can be mechanically varied; flatness of surfaces is of the order of 1/20 to 1/100 wavelength. Scanning is achieved by changing the spacing mechanically or by changing air pressure. The intensity at the ring center is recorded photo-electrically.

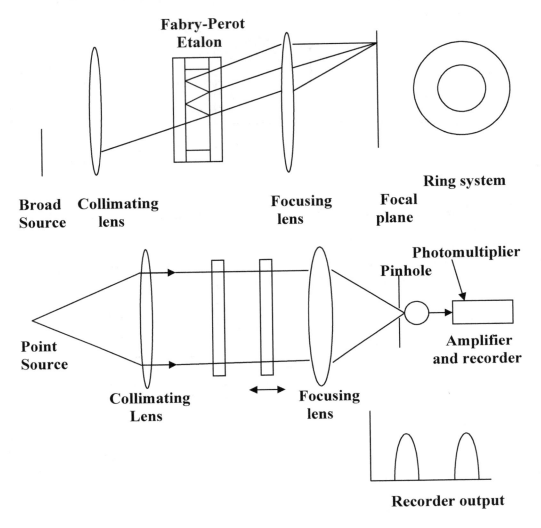

Figure-3A: Fabry-Perot Scanning Interferometer

Multiple interference has application for development of anti-reflecting and high reflecting coatings on dielectric mirrors. Band pass interference are also developed based on multiplayer interference.

DIFFRACTION

Diffraction is bending of light in shadow region of opaque object or spreading of light when it is passed through small hole or narrow slit. Diffraction phenomena can be explained qualitatively by Huygens' principle, which states propagation of a light wave can be predicated by assuming that each point of the wave can be predicated by assuming that each point of the wave front act as the source of a secondary wave that spreads out in all directions. The envelope of all the secondary wave is the new wave front.

Fraunhofer and Fresnel Diffraction

Fraunhofer diffraction occurs when both the incident and diffracted wave are effectively plane waves. **Fresnel diffraction** occurs when source or the receiving point are close to the diffracting aperture.

Multiple slits or diffraction gratings: Gratings are multiple slits of equal width and spacing.

Let the aperture consist of a grating, that is, a large number N of identical parallel slits of width b and separation h. The intensity distribution function is given by

$$I = I_0 (\sin \beta / \beta)^2 (\sin N \gamma / N \sin \gamma)^2$$

The factor N has been inserted in order to normalize the expression. This makes I $= I_0$ when $\theta = 0$

Again the single-slit factor $(\sin \beta / \beta)^2$ appears as the envelope of the diffraction pattern. Principal maxima occur within the envelope at $\gamma = n\pi$, n = 0, 1, 2, ..., that is,

$$n \lambda = h \sin \theta$$

Which is the grating formula giving the relation between wavelength and angle of diffraction? The integer n is called the order of diffraction.

Secondary maxima occur near $\gamma = 3\pi / 2N, 5\pi / 2N, \ldots\ldots$, and zeros occur at $\gamma = \pi / N, 2 \pi / N, 3 \pi / N, \ldots\ldots$, if the slits are very narrow, then the factor

$$\sin \beta / \beta = 1$$

The first few primary maxima, all have the same value, namely, I_0

Resolving Power of Gratings The angular width of a principal fringe, that is, the separation between the peak and the adjacent minimum, is found by setting the change of the quantity Nγ equal to π, that is, $\Delta\gamma = \pi / N = \frac{1}{2} kh \cos \theta \, \Delta\theta$, or

$$\Delta\theta = \gamma\lambda / Nh \cos \theta$$

Thus if N is made very large, then $\Delta\theta$ is very small, and the diffraction pattern consist of a series of sharp fringes corresponding to the different orders n = 0, ± 1, ± 2, On the other hand for a given order the dependence of θ on the wavelength gives by differentiation

$$\Delta\theta = n \, \Delta\lambda \, / \, h \cos\theta$$

This is angular separation between two spectral lines differing in wavelength by $\Delta\lambda$.

Combining above equations, we obtain the revolving power of a grating spectroscope according to the Rayleigh criterion

$$RP = \lambda \, / \, \Delta\lambda = Nn$$

In other words, the resolving power is equal to the number of grooves N multiplied by the order number n. Figure below shows Fraunhofer diffraction pattern for a many lines grating illuminated with two different wavelengths.

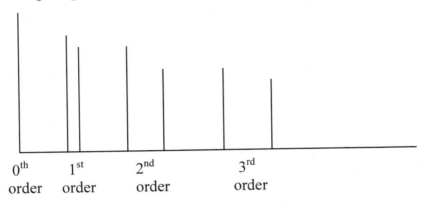

0th 1st 2nd 3rd
order   order   order   order

Diffraction gratings used for optical spectroscopy are made by ruling grooves on a transparent surface (transmission type) or on a metal surface that is reflection type. A typical grating may have 600 lines / mm ruled over a total width of 10 cm. this would give total of 60,000 lines and a theoretical resolving power of 60,000 n, where n is the order of diffraction used. In practice, resolving powers up to 90 percent of theoretical values are obtainable with good gratings. If the grooves are suitably shaped, usually of a saw tooth profile most of the diffracted light can be made to appear in one order, thus increasing the efficiency of the grating. The essential requirement is that the spacing be uniform, within a fraction of a wavelength. This places extreme requirements on the mechanical rigidity of the ruling machine. A plastic molding process can produce high-quality replica gratings. These are much less expensive than original gratings. Most of the gratings used in practical spectroscopy are of the reflection type. Reflection grating are made with the ruled surface either plane or concave. Plane grating require the use of collimating and focusing lenses or mirrors, whereas concave gratings can perform the collimating and focusing function as well as disperse the light into a spectrum.

Diffraction due to Circular Aperture

Relation for the circular aperture Intensity pattern is given by

$$I = I_0 [2J_1(\rho)/\rho]^2$$

J_1 is the Bessel function of the first kind.

The diffraction pattern consists of a bright central disc surrounded by concentric circular bands of rapidly diminishing intensity. The bright central area is known as the Airy disc. It extends to the first dark ring whose size is given by the first zero of the Bessel function. The angular radius of the first dark ring is given by

$$\sin \theta = \theta = 1.22 \, \lambda / D$$

This is valid for small values of θ. Here $D = 2R$ is the diameter of aperture.

Optical Resolution: The image of a distant point source formed at the focal point source formed at the focal plane of an optical-telescope lens or a camera lens is actually a Fraunhofer diffraction pattern for which the aperture is the lens opening. Thus the image of a composite source is a superposition of many Airy disks. If D is the diameter of lens opening, then the angular radius of an Airy disk, will be approximately $1.22 \, \lambda / D$. This is also the approximate minimum angular separation between two equal point sources such that they can be just barely resolved, because at this angular separation the central maximum of the image of one source falls on the first minimum of the other. This condition for optical resolution is known as the Rayleigh criterion.

Laser Speckle: The characteristic bright/dark speckle pattern exhibited when laser light is diffracted against a diffuse reflector.

202

The Photograph of optical components like corner cube prism, TIR prism, and beam splitter cum image inverting prism and dove prism used in some laser range finders are shown in photo-1a, 1b, and 1c. Photo-2 &3 shows large aperture parabolic mirror for aligning, sight and trans –receiver fabricated at IRDE and Fizeau interferometer for testing of laser rod and optical components at IRDE optical laboratory.

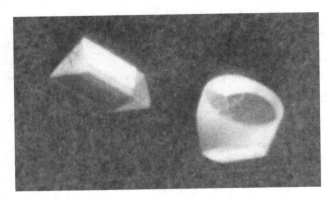

Photo-1a: Corner Cube Prism and TIR Prism

Photo-1b: Image Inverting Prism cum Beam Splitter for visible and laser light.

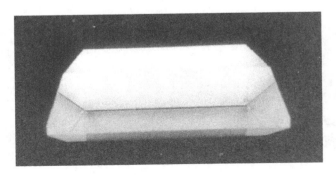

Photo-1c Dove Prism

Photo-2: Large Aperture Parabolic Mirror

Photo-3: Fizeau Interferometer

REFERENCES

1. Horne, D.F., "Optical Production Technology", Publisher: Adam Hilger, London, (1972).

2. Arthur, S. De Vany, "Master Optical techniques", Publisher: John Wiley & Sons Inc., New York, (1981).

3. The Photonics Design & Applications Handbook, Vol. 3, Laurin Publishing Co., Inc., Pittsfield, MA, USA, (1998).

4. Product Catalogue, "Optics and Optical Instruments", M/S Edmund Industrial Optics, USA, (2000).

5. Product Catalogue, "Optics Coating", M / S Acton Research Corporation", Action, MA, USA.

6. Mansharamani, N., "Design Requirements for Laser Optics", Paper presented at Workshop on Optical Design, held at IRDE, Dehradun, 13-15 Oct. 1980.

7. "Cosmetic Surface Quality", Electro-Optics Handbook, M/S Meller Griot, USA/West Germany.

8. Product Catalogue, "Optical components, M/S Litton, USA.

9. Ritter,E., "Optical coatings and thin film techniques," Laser Handbook by Arrechi, F.T . E.O. Schutz-DuBois, (North-Holland, Amsterdam 1972) pp. 897 921.

10. Baumeister,P., "Handboob of Optical Design," US Government Printing Office, Washington, DC 1963.

11. Heavens,D.S., "Optical Properties of Thin Solid Films," Butterworth, London 1955.

12. Macleod,H.A., "Thin-Film Optical Filters, 2nd Edition, MacMillan, New York (1986).

13. Vasicek,A., "Optics of Thin Films", North-Holland, Amsterdam (1960).

14. Fowlers, G.R., "Introduction to Modern Optics", second edition, Dover Publication, Inc., 31 East 2nd Street, Mineola, New York, 11501 (1975).

15. Costich,V.R.,, "Multilayer Dielectric Coatings", pp. 155-170, in Handbook of Lasers by Robert J. Pressley (Editor), Publisher CRC Press, Inc, Cleveland, Ohio, USA (1971).

CHAPTER-9

LOW-LEVEL PHOTO DETECTION

9.1 HISTORICAL BACKGROUND

In 1887, Hertz observed that ultra-violet light falling on the electrode of a spark gap caused to jump greater distance than it was left in dark. In 1888 Hollwacks found that negative charged metal surface lose their charge on exposure to ultra-violet light. It was found that this is due to emission of electrons from the surface of plate. The phenomenon of emission of electrons from the surface of metal under action of light is called photoemission and current is called photo current. On study of photoemission experimentally by Millikan, it was found that emission of photoelectrons is independent of temperature. It depends on frequency of light, and disappears at a certain frequency called the threshold frequency, depending on the particular metal used. The rate of emission depends on intensity of light, pressure and nature of gas surrounding the metal. The velocity of emitted electrons depends on frequency of light.

To explain the photoelectric effect, Einstein (1905) made use of concept of a quantum of energy, a concept which was first introduced into physics by Plank; in order to explain the distribution of energy among various wavelengths in the radiation from a "Black-Body" at temperature T. According to Plank's theory, whenever radiation is emitted or absorbed by such a body, the energy is emitted or absorbed in whole quanta, where a quanta of energy is given by

$$\varepsilon = h\nu \qquad \qquad \dots\dots\dots[1]$$

In which ν is the frequency of radiation and h is the Plank's constant. Such a quantum of energy has since received the name photon.

In Einstein's explanation of photoelectric effect, the entire energy of a photon is transferred to a single electron in the metal. Electrons in two ways use this photon of energy:

It does certain work w in dragging electron out of the atoms. Rest of this energy remains as Kinetic Energy (KE) of electron, as v is the velocity of electron emitted.

$$\text{i.e.} \quad KE = \tfrac{1}{2}\, m\, v^2 \qquad \qquad \dots\ \dots[2]$$

$$\text{Hence } \varepsilon = h\nu = w + \tfrac{1}{2}\, m\, v^2 \qquad \qquad \dots\dots\dots\dots[3]$$

$$\text{or } \tfrac{1}{2}\, m\, v^2 = h\nu - w \qquad \qquad \dots\dots\dots\dots[4]$$

$$\text{And } w = h\nu_0 \qquad \qquad \dots\dots\ \dots..[5]$$

v_0 is known as threshold frequency, w is called work function of metal and is of the order of few electron volts for most of conductors. The lowest work function was found in alkali metals, in which emission is possible when metal are exposed to visible light.

Soon based on photoelectric effect, photocell and photo-multipliers developed to detect light in visible and ultraviolet radiation. With the development of ruby laser at 0.6943 μm and He-Ne laser operating at 0.6328 μm and requirements of laser range finders for defense and survey purpose, low level detection in laser receiver was carried by photo-multiplier tubes with S-20 photo cathode using multi alkali elements as cathode material. This material has quantum efficiency of only 2.6% at ruby laser wavelength. Soon after development of more efficient neodymium lasers operating at 1.06 μm, photo-multiplier tube with S-1 response using Ag-O-Cs could not be used, as its quantum efficiency at neodymium laser wavelength was 0.01%. The p-n junction photo diode available with quantum efficiency of 25% at 1.06 μm could not be used as detection capability was limited by preamplifier noise.

9.2 INTRODUCTION

With the requirement of low noise and efficient detector for neodymium laser wavelength, fiber optical communication with low dispersion effect at 1.3 to 1.5 μm, image intensifier tubes for night vision viewing devices for military application, details understanding of photo emission and photo detection was carried out in mid sixties. This has resulted in better coupling of light in photosensitive materials, deposition of alkali metals and its oxide on negative affinity materials for better quantum efficiency and improving its response towards infrared wavelength. Introduction of hetero structures in semiconductors, low band gap semiconductors, internal carrier multiplying region in semiconductors has resulted in development of high quantum efficiency photo detectors, avalanche photodiodes and fast response cooled photo diodes for detection of far infrared light. In this chapter, we shall confine for detectors used for laser range finders i.e. photo-multipliers tubes for visible, silicon photo and avalanche, InGaAs photo diodes and avalanche diodes for eye safe laser at 1.55 μm and cooled fast HgCdTe detectors for carbon dioxide lasers in range applications.

Development is going on for tuned quantum well semiconductor detectors, with optimum quantum efficiency at any laser wavelength, for background noise reduction, but still these detectors are costly and on experimental stage and are not commonly used in range finders.

9.3 TERMINOLOGY

The parameters of photo detectors are defined by the following terms

9.3.1 Responsivity (R) The responsivity of detector is a measure of the optical signal voltage or current developed by the detector when exposed to known amount of optical power. It is usually expressed as volts/watt or amperes/watt.

9.3.2 Detector D* It is a figure of merit (pronounced Dee Star) and is measure of S/N ratio normalized to unit noise bandwidth and not detector area i.e.

$$D^* = \frac{\text{(Detector sensitivity area)}}{\text{NEP}} \; Hz^{1/2} / \text{watt} \qquad \dots\dots\dots[6]$$

9.3.3 Noise Equivalent Power (NEP) The NEP is a measure of the minimum power that can be detected. It indicates the power required to generate a signal to noise ratio of 1, when the detector noise is referred to one cycle bandwidth.

$$NEP = \frac{P_0 \, A_d \, V_n}{V_s \, (\Delta f)^{1/2}} \; \text{watts/Hz} \qquad \dots\dots\dots\dots[7]$$

Where P_0 = watts/cm^2.
A_d = Area of detector surface
Δf = Noise Bandwidth
V_s = Signal Voltage.

9.3.4 Quantum Efficiency (η): Quantum efficiency is related to responsivity i.e.

$$R = q \, \eta / \, h\nu \qquad \dots\dots\dots[8]$$

where q = 1.602 x 10^{-19} coulombs.

9.3.5 Time Constant: The detector time constant is the finite time taken to respond a signal?

9.3.6 Spectral Response: It is a response of detector at various wavelengths.

9.4 THE PHOTOMULTIPLIER TUBES

The photo-multiplier tubes (PMT) are generally used for laser range finders operating in visible radiation i.e. Nd:YAG-second harmonics (0.532 μm), He-Ne (0.6328 μm) and ruby (0.6943 μm).

It consist of photo cathode [1], made of one of the photo emitting material illustrated in table I, several secondary-emitting surfaces called dynodes and a collector anode. A progressively increasing voltage is applied to the different dynodes, usually through a resistance divider network. Electrons photo emitted from the cathode are guided and accelerated towards the first dynode, where for each incident electron more than one electron is released. The electrons emitted at the first dynode impinge upon the second dynode and their number is again increased, this process continues until the electron is collected at the anode to be registered as an output signal. This process takes time of 10 to 100 nano seconds. The electron multiplication from dynodes of PMT is of the order of 10^6 to 10^7.

The material of dynode is chosen that it should give good secondary emission of electrons even at low accelerating voltage i.e about 100V. The dynodes coated with Cs_3Sb gives maximum secondary emission, but have maximum operating temperature of 75^0C. Generally alloy of Ag-Mg or Cu-Be are used, these alloys with 2% oxide of Mg/Al or Be respectively gives good yield of secondary electrons and can be used at high temperature and also have low themoionic emission. Dynodes coated with negative affinity material i.e. GaP(Cs) gives good yield of secondary electrons is also used for compact, low noise PMT and is used as 1st dynode. A PMT is shown in figure -1, illustrating its various essential parts.

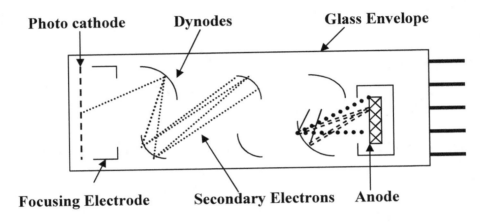

Figure-1 Schematic of Photo Multiplier Tube (PMT)

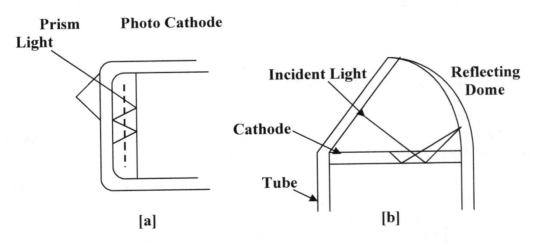

Figure-2 Photo Cathode Modified for more Coupling of Light

In PMT, generally semitransparent cathode are deposited on the inside of glass wall of tube, more than 50% of light is coupled out of photo cathode, without causing electron emission, especially on longer wavelength side of spectrum. Therefore to increase quantum efficiency and extend the response to higher wavelength side, either thickness of cathode layer is increased or light is coupled into cathode that it gets reflected back into photo sensitive material as shown in figure-2a and 2b. The spectral response of cathode materials as given in Table - I is shown in figure-4.

Now days, instead of dynodes in PMT, channel electron multipliers can be used, these multipliers improves performance of photo-multipliers. The viewing devices, with gated channel multiplier tube, can be used for illumination and range gated viewing of target using pulse laser under adverse weather conditions. In simple form, the channel multiplier is a straight glass capillary tube with length to diameter ratio of 50 to 100. The inside diameter is between 0.1 to 1 mm. On the inside surface of glass tube, a layer of semiconducting material such as a metallic oxide, having secondary emission characteristics suitable for the electron multiplication process is deposited. When voltage is applied across the ends of the tube on evaporated contact of the resistive material on the ends of tube, an electric field is established down the length of the tube. Any electron emitted from the inside wall of the tube will be accelerated down the tube. Because electrons are emitted with finite later velocities i.e. energy of few electron volts, they will drift across the tube, intercepting walls of the channel tube, where electron multiplication

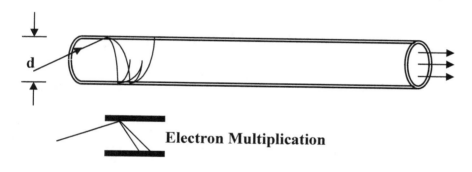

Electron Multiplication

Figure-3 PMT with Continuous Channel Multiplier

TABLE-I

Photo emissive Material	Response Type	Spectral Range in μm	Wavelength of Maximum Response λ_{max}	Response mA / Watt	Quantum Efficiency at λ_{max}
1. Ag-O-Cs	S-1	0.3 to 1.1	0.8	3	2
2. (Cs)NaKSb	S-20	0.3 to 0.8	0.4	80	-
3. (Cs)Na$_2$KSb	S-25	0.3 to 0.9	0.4	40	-
4.GaAsP(Cs)	GEN. III	0.25 to 0.7	0.4	-	40%
5.GaAs(Cs)	GEN. II	0.4 to 0.85	0.8	-	12.5%

occurs. When the applied voltage and tube dimension are such that electron gain sufficient energy between encounters with the wall, such that, on the average, more electrons introduces at the input of device, results in an electron cascade. The process of electron multiplication is shown in figure-3. In these devices gain of 10^5 to 10^7 in electron multiplication can be easily achieved. The dark noise and dark current of a continuous channel photo-multiplier, is less than photo-multiplier tube with discrete dynodes.

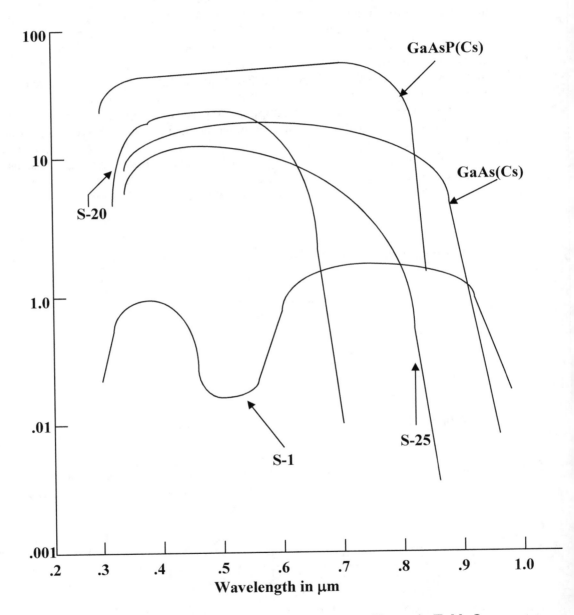

Figure-4 Spectral Response of Cathode Material Shown in Table-I

9.4.1 Sources of Noise in PMT

The main sources of noise in photo multiplier are due to dark current. The dark current arises due to thermal emission of electron from cathode surface or dynode material; leakage surface current over glass surface of tube between various leads due to high voltage applied, ionization of residual gas molecules in tube due to accelerated electrons, Electro luminescence of glass envelope, cosmic rays, radioactive substances or EMI from external fields. Other source of noise is due to shot noise in background illumination. Noise due to Johnson noise, Flicker noise or preamplifier noises have negligible effect on detection capability of PMT. The ratio of root mean square (rms) signal voltage S developed across anode resistance to noise voltage N due to shot noise due to PMT dark current I_d, background illumination and secondary electron emission and its average multiplication gain g per dynode is given [1] by relation

$$\frac{S}{N} = \frac{q\,\eta(\lambda_s)\,m\,F}{[qB\{\,I_d + q\eta(\lambda_s)\,F\,m + q\,\phi_B(\lambda)F\eta(\lambda)d\lambda\}\,E_x + 2kTB/g^{2n}R_L]} \qquad ....[9]$$

Where B = Bandwidth in Hz
  q = 1.602×10^{-19} coulombs
  $\eta(\lambda_s)$ = Quantum efficiency of cathode at signal wavelength λ_s
  F = First dynode electron collection factor
  m = modulation index of signal
  $\phi_B(\lambda)$ = Background photon flux at cathode in photons/second at wavelength λ
  $\eta(\lambda)$ = Quantum efficiency of cathode at wavelength λ
  E_x = Excess noise factor due to secondary emission from dynode
  k = Boltzmann Constant
  T = Absolute temperature
  n = number of dynodes in PMT.

E_x, excess noise factor depends upon the gain of 1st dynode and subsequent dynodes, i.e.

$$E_x = 1 + 1/g_1 + 1/g_1 g + 1/g_1 g^2 + \ldots\ldots + 1/g_1 g^{n-1} \qquad \ldots\ldots [10]$$

g_1 is gain of 1st dynode, high voltage is applied between cathode and 1st dynode, for large electron emission to keep this excess noise factor almost approaching 1.

It has been observed experimentally that detection capability with PMT type 7265 with S-20 response of M/S RCA, a ruby laser echo pulse of 20 nano second duration with peak power of 10^{-10} watts can be easily detected in direct detection mode. The detection capability is limited by background noise due to sun even in a highly directive ranging receiver using narrow interference filter of bandwidth of 20 A^0.

9.5 PHOTOELECTRIC EFFECT – SEMICONDUCTOR

Light with photon energy greater is absorbed in semiconductor having band gap width i.e. difference in energy between minimums of conduction band energy to maximum of valance band of semiconductor, generation of free carriers takes place in the semiconductor. This effect is known as **Photoelectric Effect** in semiconductor.

The band gap energy varies from semiconductor to semiconductor and therefore cut off wavelength λ_c for photoelectric effect varies from semiconductor to semiconductor and is given by relation

$\lambda_c = 1.24/ E_g$ μm ....[11]

Where E_g = Band gap energy of semiconductor in electron volts (ev)

Table-II Shows band gap energy and cut off wavelength of various direct and indirect semiconductor materials.

Table – II Band Gap Energy (E_g) and Cut of Wavelength

Semiconductor	Type	E_g in eV	Cut off Wavelength in μm
$Hg_{0.8} Cd_{0.2} Te$	Direct	0.1*	12
InAs	Direct	0.35	3.5
GaSb	Direct	0.67	1.85
Ge	Indirect	0.72	1.70
Si	Indirect	1.10	1.20
InP	Direct	1.26	1.00
GaAs	Direct	1.35	0.90
CdTe	Direct	1.4	0.89
AlSb	Direct	1.6	0.78
CdSe	Direct	1.7	0.73
GP	Direct	2.24	0.55
CdS	Direct	2.50	0.50
ZnSe	Direct	2.60	0.48
ZnS	Direct	3.70	0.34

* at 77^0 Kelvin (K)

9.5.1 Direct & Indirect Semiconductors

There are two types of semiconductors- (a) Direct and (b) Indirect.

In direct semiconductors phase of electron in maximum of valence band is same as in minimum of conduction band as illustrated in Figure-5a. While in indirect semiconductor, phase of conduction electron is not same as in valence band as shown in figure-5b. Therefore transition of electron from valence band to conduction band or vice-versa due to absorption or emission of photon takes place along the absorption or emission on phonon (lattice vibration in semiconductor crystal). The probability of absorption therefore is less in indirect semiconductor is less as compared to direct semiconductor.

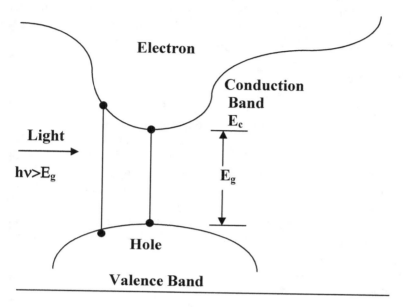

Crystal Momentum
[a]

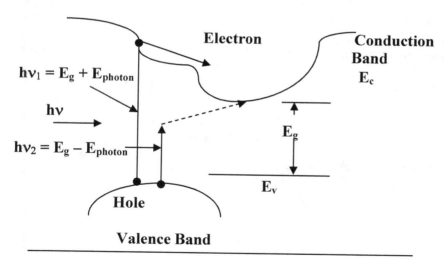

Crystal Momentum
[b]

Figure-5 Energy Phase Diagram of [a] Direct and [b] Indirect Semiconductor

214

9.5.2 Photovoltaic and Photoconduction Mode

There are two types of modes of operation in semiconductor with p-n junction. The carriers generated in depletion layer of p-n junction due to absorption of light in depletion layer can cause voltage diffusion across the junction, if no voltage bias is applied; this mode of operation is known as photovoltaic mode detection of light. If p-n junction is reverse bias, due to absorption of photon in depletion layer can increase its conductivity. This mode of operation is known as photoconduction mode.

9.5.3 Quantum Efficiency

The ratio of number of carriers due to absorption of single photon in a semiconductor is known as its quantum efficiency. The quantum efficiency depends upon coupling factor of light in semiconductor, absorption coefficient and width of depletion region. The dielectric constant of semiconductor material is 3.5, which gives 30% reflection loss of incident radiation. Therefore with antireflection coatings on semiconductor surface this loss can be reduced. The direct band gap semiconductors have large absorption coefficient, thus the material i.e. alloy of group III-IV are used for detector. The width of depletion region can be increased, either by edge illumination or by incorporating wide intrinsic region-i between -p and –n junction. This type of photodiode is called p-i-n photodiode.

9.5.4 p-i-n Photodiode

The p-i-n photodiodes are generally used for low level detection of light due to fast response, low dark current / low noise level and high quantum efficiency. The constructional details of a typical Si and InGaAs p-i-n with guard ring structure used for low level detection of light is shown in figure-6a and –6b. The guard ring structure reduces current due to imperfections in surface and also edge breakdown voltage. The electrons and holes generated in wide i-region drifts quickly towards n and p region due to reverse voltage, contributing conduction of current in the device. The intrinsic layer is undoped or very lightly doped. Concentration in very lightly doped semiconductor (n⁻

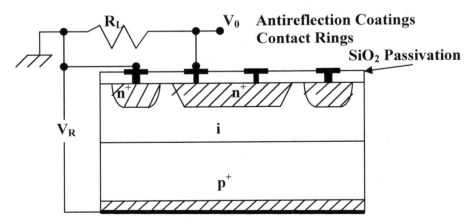

Figure-6a Silicon p-i-n Photodiode with Guard Ring

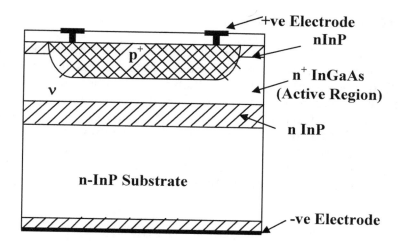

Figure-6b In GaAs/InP Planar Type (Hetero Junction) p-i-n Photo Diode

intrinsic layer called ν - layer or p⁻ layer which is called π - layer) is between 10^{13} to 10^{16} ions / cm³. Heavily doped n⁺ or p⁺ concentration is 10^{18} ions per cm³

9.5.5 Photodiode Wave-Guide Type

Wave-Guide type of photodiodes is a heterostructure photodiode in which thin long active i- region is of high index material and is surrounded by heavily doped n⁺ and p⁺, low refractive index material. Light coupled through ends as shown in figure-7, travels to a longer distance in active region i.e. due to wave-guide structure. Even at low reverse voltage, transit time of photo generated carriers is smaller with the result that this type of photodiode has cut off frequency of more than 50 G.Hz. and quantum efficiency more than 70% in which wavelength antireflection coatings are given.

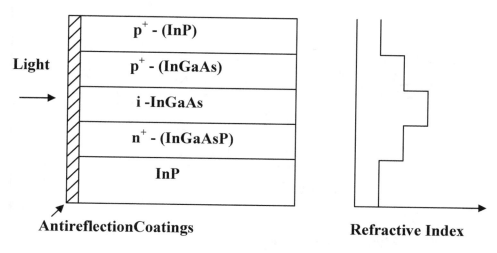

Figure-7 Wave Guide Type (Hetero Structure) p-i-n Photodiode

9.5.6 Noise in Photodiodes

Noise in photodiodes depends on dark current fluctuations, background shot noise and thermal noise in resistances. The dark current in semiconductor is due to imperfections on surface and crystal resulting in surface and bulk currents in device. These currents can be reduced by use of highly pure semiconductor free from crystal defects. The use of guard ring reduces surface currents, as surfaces of semiconductor have more imperfections. The bulk currents due to diffusion currents, narrow band gap and carrier generation-recombination can be reduced by cooling the device. In case of direct semiconductor with thin active region with high reverse voltage, tunneling of carriers contribute to dark currents also. The signal to noise ratio for photodiode is given by relation

$$S / N = I_s^2 / [2q (I_d + I_B + I_s) B + 4 k T B / R_{eq}] \qquad \ldots\ldots\ldots\ldots[12]$$

Where I_d is dark current
I_B = Current due to back ground illumination
I_s = signal DC current
B = Band width in Hz.
q = 1.602 x 10^{-19} Coulombs
T = Operating temperature in Kelvin
k = Boltzmann constant
R_{eq} = Equivalent resistance of device including load resistance.

9.5.7 AVALANCHE PHOTODIODES

Avalanche photodiodes (APD) are used in laser range finder for detection of laser echo's from targets. Due to internal multiplication and high quantum efficiency at laser wavelength, the detection capability is limited by back ground noise due to illumination by sun during daytime. Like heterostructure p-i-n, APD consist of four region as

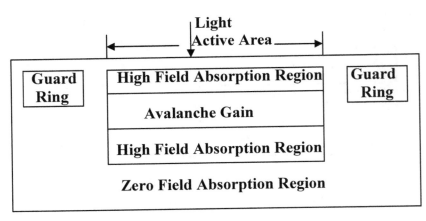

Figure-8 Schematic of Avalanche Region

illustrated in figure-8. Most commonly type of APD used for ranging applications are Si reach through type APD and InGaAs / InP planar type APD.

9.5.7.1 Silicon Reach Through Type APD

In reach through structure, changing the thickness of intrinsic active region can accurately control quantum efficiency at long wavelength and speed of response. High speed, high gain and low noise is obtained by separating the depletion region in two, a wide drift region in which photons are absorbed and a narrow multiplying region in which photo generated carriers are multiplied. This type of detector is shown in figure-9

The p and n diffusion making the avalanche region are carried out in sequence and adjusted so that when in reverse bias, the depletion layer of the diode just reaches through π (high resistively p-type) region. With peak electric field is 5 to 10% less than the required for avalanche.

Additional applied voltage causes the depletion to increase the depletion to increase rapidly to p^+ contact, while the field through out the device increases slowly. The π–region is about 5000 Ω and giving 200µm depletion width at < 100V over that require to achieve reach-through. In all cases the device is operated in the fully depleted mode and the carriers are collected by drift only. Optimum condition occurs when light wavelength λ is absorbed in π region i.e. through p^+ contact. The electron generated in π region swift to high field region for multiplication. Holes produced there, traverse to π region in p^+ contact.

The avalanche voltage varies with temperature; variation of avalanche voltage with temperature for 100 µm wide depletion region is shown in figure-10. For constant gain, voltage should vary by 1.8V / ^{0}C.

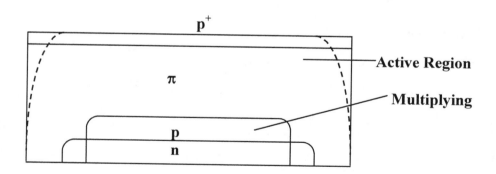

Figure-9 Silicon Reach Through Avalanche Photodiode

218

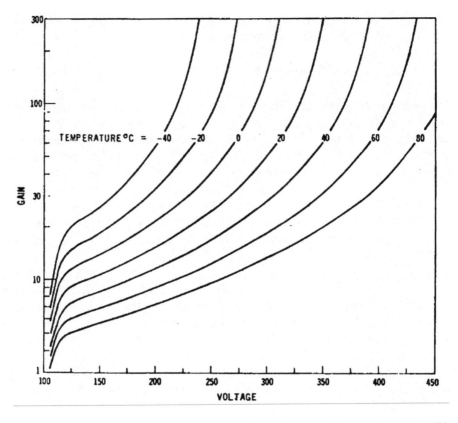

Figure-10 Variation of Avalanche Voltage for a Typical Reach Through Photodiode

9.5.7.2 InGaAs / InP Planar Type APD

InGaAs / InP type with guard ring and planar structure is illustrated in figure-11.

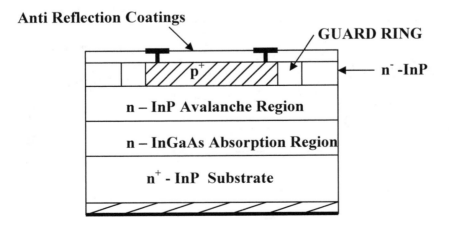

Figure-11 Planar Type InGaAs / InP Avalanche Photodiode with Guard Ring

It consists of active n – InGaAs absorbing region, where carriers are generated. These carriers are accelerated in high field adjacent region, where multiplication of carriers takes place. The multiplying region and active region are separated for reduced noise and high quantum efficiency. The quantum efficiency is more than 70% for Er:Glass laser wavelength. To avoid edge break down, a guard ring is formed by diffusion or ion-implantation. The carrier multiplication gain can be 50 to 500 depending upon operating voltage.

The quantum efficiency curve with wavelength for Si and InGaAs / InP photo detectors are shown in figure-12.

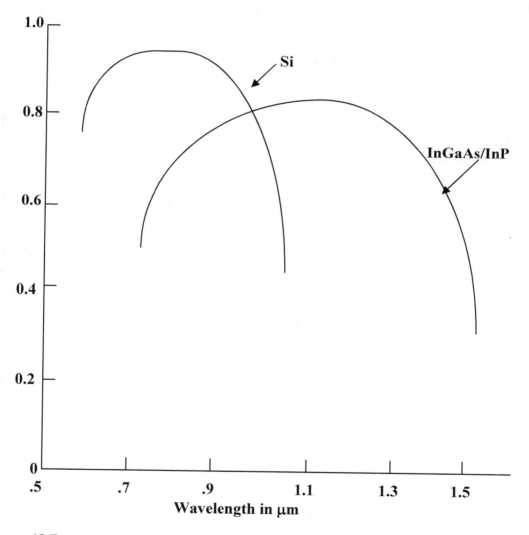

Figure-12 Responsively Curve of Si and InGaAs/InP Photodiode with Antireflection Coatings

9.5.7.3 Noise in APD

The noise sources in APD are similar to photodiodes. Here all photo generated and dark current, except surface leakage currents gets multiplied along with signal. Due to high quantum efficiency and internal carrier multiplication, Johnson noise due to equivalent resistance, and drag current have negligible effect of noise power level, except shot noise due to background illumination and excess noise generated due to ionization of carriers i.e. holes and electrons. This excess noise factor due to carrier's ionization is given by

For electron injected into high field region at x = 0, excess noise factor for electron F_n is

$$F_n = M_n \{ 1 - (1 - k)[(M_n - 1) / M_n]^2 \} \qquad \ldots\ldots\ldots[13]$$

For holes injected at x = w, the excess noise factor is given by

$$F_p = M_p [1 - (1 - 1 / k)] [(M_p - 1) / M_p]^2 \qquad \ldots\ldots\ldots[14]$$

Where $k = \beta_p / \alpha_n$, ratio of ionization coefficients of holes to electrons and M_n and M_p are electron and hole multiplication factors.

For k = 1, noise power generated by detector varies as M^3 , noise will reduce for small value of k.

For k = 1 / 20, value of noise power vary as $M^{1.15}$. Figure-13 shows excess noise factor verses gain for various values of k – ratio of ionization coefficients of holes to electrons.

Following relation gives signal to noise ratio of an avalanche photo diode

$$S / N = (i_s / I_n)^2 = \frac{(P_0\, m\, R_0\, M)^2 / 2}{2q\, B[I_{ds} + (P_0\, R_0 + I_b + I_{db})\, F\, M^2] + i_{na}^2} \qquad \ldots\ldots[15]$$

Where P_0 = average value of intensity of light.
m = modulation depth of light.
M = Gain of APD.
q = Electron charge i.e. 1.602×10^{-19} coulombs.
B = Bandwidth in Hz.
I_{ds} = Surface dark current of device (APD).
I_{db} = Bulk dark current of device (APD)
I_b = Photocurrent due to background illumination
M = Gain of APD
F = Excess noise factor

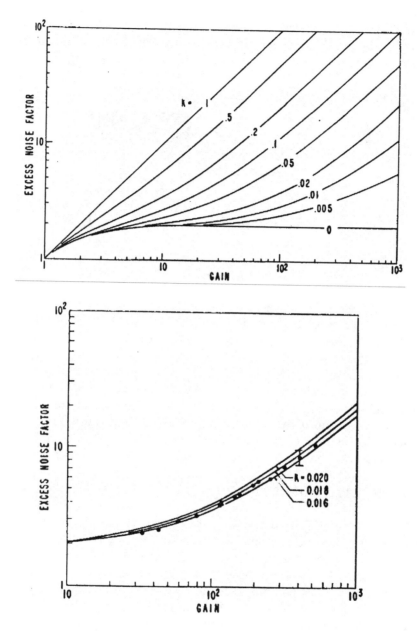

Figure-13 Excess Noise Factor for Silicon Reach through APD (Courtesy of RCA Review)

9.5.8 INDIUM ANTIMONIDE (InSb) DETECTOR

InSb is the first III-V group material with narrow band gap developed in 1950 to be used for infrared detectors. Detectors using this material have peak response at 5 μm when operated at 77^0K and this material can be prepared into a pure single form using

222

convention techniques. Epitaxy methods such as liquid phase epitaxy (LPE), molecular beam epitaxy (MBE) or chemical vapor deposition (MOCVD) are used for more sophisticated structures of modern InSb devices. A photosensitive area of the detector is made from several tens of μm^2 to several mm^2, depending upon usage.

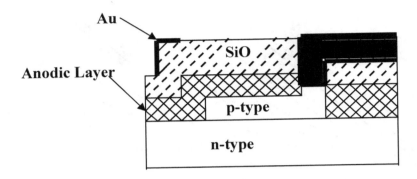

Figure-14: Indium Antimonide (InSb) Detector

Generally for Lidar / ranging applications detectors with 0.25 mm^2 to 1 mm^2 photosensitive area are used. Small area detectors have fast response time. For 300^0 K operating device has optimum thickness of 5 – 10 μm and for 195^0 to 77^0 K element thickness is 25 μm. Electrical contacts are applied by soldering using indium based contact material. For long life operation, passivation, layer approximately 50 nm thick is obtained by anodizing usually up 0.1 N KOH solutions. Additionally, on top of the passivation layer of SiO or ZnS about 0.5 μm thick is evaporated to provide a more stable surface. For n-type of substrate, Zn or Cd impurities are doped using liquid phase epitaxy or ion implantation and donor concentration is in the range of $10^{14} - 10^{15}$ cm^{-3} at 77^0 K. InSb detector is shown in Figure-14. The spectral response at 77^0 K is shown in figure-15

9.5.9 MERCURY CADMIUM TELLURIDE (MCT) DETECTORS

MCT i.e. $Hg_{1-x}Cd_xTe$ detectors also known as thermal quantum detector are commonly used for detection of mid or far infrared radiation unto 30 μm by cooling it with liquid helium temperature. By changing value of x, its peak response can be changed to any wavelength. It has peak response for x = .2 at temperature of liquid nitrogen (77^0 K). These detectors are fast and more sensitive as compared to thermistor, bolometers or pyroelectric. MCT detectors are direct band gap semiconductors having narrow band gap of order of 0.1eV, the energy of impurity or dopant atoms lies in forbidden gap of semiconductor. Therefore these semiconductors are known as extrinsic semiconductors. Due to excess carrier photo generation recombination noise, when used with reverse bias i.e. photo conductive mode, their response is fast even at photovoltaic mode of operation. Therefore these detectors are used without any external bias. In this mode of operation photo generation recombination noise is negligible. The spectral response of cooled detector is shown in figure-15 and the constructional details of most commonly type of photodiode are illustrated in figure –16a and –16b

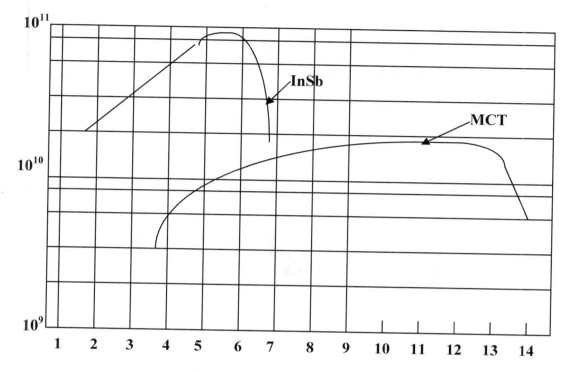

Figure-15 Spectral Response of InSb and MCT at 77^0 K

Table-III Specifications of Detectors Used in Laser Range Finders

	Si p-i-n Photodiode YAG-100	Si APD C 30954E	MCT
Active Area	5.1 mm^2	0.5 mm^2	120 μm dia.
Spectral Response	0.35 – 1.13 μm	0.35 – 1.13 μm	5 – 14 μm
Responsivity	0.65 A / W at 0.9 μm	36 A / W at 1.06 μm	-
Quantum Efficiency	38 % at 1.06 μm	-	36 % at 10.6 μm
D* (W^{-1} cm Hz^{-1})	1×10^{12}	-	2.1×10^{10}
Noise Current	-	2.2×10^{-13} A Hz^{-1}	-
Dark Current	20 nA	50 nA	-
Operating Voltage	180 V	Typ. 297V (at 22^0C)	-
Breakdown Voltage	-	Typ. 351V	-
Operating Temperature	-40 to + 70^0 C	-40 to + 70^0 C	77^0 K
Manufacturer	E.G. & G.	RCA	SAT

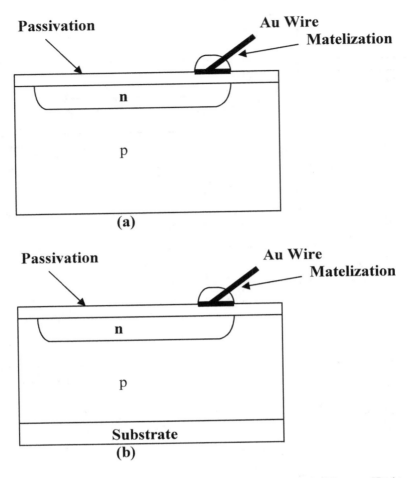

Figure-16 (a) MCT wit Planar Bulk Base and (b) Planar Epitaxial

SUMMARY

In this chapter, commonly used detectors for laser range finders or distances measuring are described. The detectors described are photo multiplier tubes, silicon pin and avalanche photodiodes, InGaAs-InP pin and avalanche photodiode for low level detection of signal in visible and near infrared spectrum of light. Cooled MCT photodiode described is use for TEA-CO_2 laser range finder. The choice of detector does not depend only for quantum efficiency, but also on noise figure and gain. For example InGaAs photo diode have more quantum efficiency for Nd:YAG laser, but silicon with less quantum efficiency is used due to less noise figure and cost. In photon thermal detector MCT is described, though low level detection becomes costly and complex due to cooling, other type of detectors cannot be used due to slow response and very low detectivity at CO_2 wavelength. Table-III shows specification of some of the commercial detectors used for ranging application.

REFERENCES

1. Marton, L., "Advances in Electronics & Electron Physics", Academic Press, London (1973).

2. Mitsue Fukuda, "Optical semiconductor Devices", Publisher: John Wiley & Sons Inc., New York (1999).

3. Webb, P.P., McIntrye, R.J.and Conradi, J., "Properties of Avalanche Photodiodes", RCA Review, Vol.35, June 1974, pp. 234-277.

4. Mansharamani, N., "Use of Photo detector in Laser Receivers", unpublished report, Royal Signal & Radar Establishment, U.K. (1977).

5. Antoni Rogalski, "Infrared Photon Detectors", SPIE Optical Engineering Press, Washington, USA (1995).

6. Gaussorgues, G. "Infrared Thermography", Chapman & Hall, London (1994).

7. RCA-Electro-Optic Handbook-10 (1968).

8. Product Catalog, "Photon Devices", EG&G Inc. USA (1985).

9. Koren, B. and Szawlowski, M., "Large Area Avalanche Photodiodes Challenges PMT's", Laser Focus World, vol. 34, No. 11, November 1998, pp. 71-80.

10. Kenneth, K., "Photo-multipliers Lead in Low-Light Detection', Laser Focus World, Vol. 35, no. 10, October 1999, pp. 99-103.

11. Muller, J., "Photodiodes in Optical Communication", Electronics and Electron Physics, Vol. 55, pp. 189-308, Academic Press (1981).

12. Iyer,R.S., "Threshold Detection in Pulsed Laser Ranging", Appl. Optics (USA), Vol. 17, No. 7, June 1976, pp. 1349-50.

13. Keyes,R.J. (Editor), "Optics and Infrared Detectors", Vol. 19, Topics in Applied Physics, Springer-Verlag, Berlin (1977).

14. Burd,A.M., Leichenko,Yu.A.and Motenko,,B.N., "Design of Pulsed Signal PhotodetectorUsing an Avalanche Photodiode", Soviet J. of Optical Technology, Vol. 42, No. 11, November 1975, pp. 638-640.

15. McKinley,V. and Monds,F.C. "Unsupervised Learning and Detection of Pulse Signals", The Radio & Electronic Engineers, Vol. 44, No. 11, November 1974, pp. 583-92.

16. Koramov,A.A. and Bhemol,D.M., "Optimal Extraction of Optical Signals from an Interference background in the Presence of Internal Noise at the Radiation Receiver", Autom. & Remote Control (USA), Vol. 35, No. 9, September 1974, pp. 1413-1423.

17. Kurhap,A.C., "Background Canceling Optical Detection System", Patent USA 3430047, 4 January 1965, Published 25 February 1969 USA 423282.

18. Persky,D.E., "New Photo Multiplier Detector for Laser Application", Laser Journal, Vol.1, No.1, November/December 1969, pp. 21-23.

19. Smith,R.A., "Detectors for Ultraviolet, Visible and Infrared Radiation", Appl. Optics (USA), Vol. 4, No. 5, June 1965, pp. 631-638.

20. Anderson,L.K. and McMurty, B.J., "High Speed Photo Detector", Appl. Optics (USA),Vol. 5, No. 10, October 1966, pp. 1573-1587.

21. Fisher,M.B., and McKenzie, "A Travelling-Wave Photo Multiplier" IEEE J. Quantum Electronics, Vol. QE-2, August 1966, pp. 322-327.

CHAPTER-10

LASER RANGING SYSTEMS

10.1 INTRODUCTION

Laser ranging systems / range finders [13] and distance measuring equipment developed within two years after demonstration of ruby laser by Maiman [1] were used for ranging military targets and accurate distance measurement for survey purpose. Within the span of forty yeas of development of different laser sources and low level detection: The ranging systems and lidar are finding wide application in meteorology [2], air pollution monitoring and pollution control [3-5], remote detection [7-11], collision avoidance [8] range gated illumination and ranging under adverse weather conditions [6], air-craft safe landing under sever atmospheric turbulence and poor visibility. For military role targets are armored vehicles, aircraft, ships, military units, which are called uncooperative targets where reflection of light is diffuse and its reflectivity coefficient depends on nature of targets. While for space role or survey purpose retro-reflector can be used at target. Therefore for military application high power lasers are used and maximum range can be measurement only for few tens of kilometers, while in case of cooperative targets range can be from few kilometers for ground targets, to hundreds of thousand kilometers for space applications. Unlike microwave radar, laser beam is highly directive and can be used to measure range for small targets on ground. Accurate ranging help in accurate aiming of ammunition to destroy military targets, which help in operation time and money saving. For survey purpose, high accurate range measurement can be carried within accuracy of mm. Range finders in space can precisely measure land drift to predict earthquakes [12].

For ranging military targets, diffuse laser light is detected from enemy targets, for survey purpose and space application, reflected light from retro-reflector is detected, for remote detection, air pollution, investigation on scatter light spectrum or differential attenuation measurement at close two wavelength of light is carried out. For meteorology, scattering, beam attenuation and beam divergence effects are measured using lidar. It is not possible to cover all aspects of ranging systems / instruments, here only range finger for military application and survey purpose are described.

First portable laser range finder was developed [13] by U.S. Army Electronics Research and Development Laboratory (USAELRD) in November, 1961 using rotating prism Q-switched ruby laser source with peak power of 2 MW. In receiver photo multiplier with narrow band interference filter used to filter out background light and range of uncooperative ground targets measured upto 10 km. Since then many type of laser range finders developed in various research laboratories and produced. The laser range finders developed for ranging non-cooperative targets are pulsed ruby, Nd doped in silicate, phosphate and Yttrium Aluminum Garnet (YAG), Er:Glass, Holmium, CO_2 both pulsed, highly stable continuous wave (cw) lasers are commonly used. Most of the systems use Nd:YAG Q-Switched laser source-due to better stability, efficient pumping efficiency both with Xenon flash lamp and diode sources, good thermal conductivity and availability of efficient and low noise silicon APD with good performance and operation at normal atmospheric temperature. Because of high cost and less shelf life and detector operation at liquid nitrogen temperature, there is little demand of CO_2 laser range finders and are rarely used with thermal sight under dusty field conditions.

Er:Glass laser, now a days commonly used for ranging military targets, as its wavelength is safer to eye [14]. Highly directive He-Ne lasers in cw-mode or semiconductor GaAs pulse lasers are being used for distance measurement of cooperative targets to accuracy within mm [15].

In this chapter laser range finder developed at IRDE in various configurations are described [16]. Design aspect, range equation, background noise on detector due to sun illumination on target, assembly of a range finder, alignment procedure and testing are also given. Ranging systems using continuous wave He-Ne lasers and pulse GaAs lasers are also described for survey purpose.

10.2 THE BASIC PRINCIPLE

The basic principle on which the laser range finders or distance-measuring instrument is based on measurement of to and fro travel time of light between range finder and target. If d is the distance of target from range finder and n is average refractive index of air for laser wavelength used, time t of travel of light for to and fro path is given by relation

$$t = 2 d n / c \qquad \qquad \dots\dots\dots[1]$$

Range accuracy depends on how precisely time t is measured.

10.3 RANGE EQUATION

Optical power P_r (watts) received at photo detector of laser receiver with aperture A_r (meter square) from an uncooperative target of area A_t (meter square) having diffuse reflectivity ρ, at range R (meter) when laser falls normally at target is given by

$$P_r = \frac{\rho . P_t . A_r}{\pi R^2} . \frac{4 . A_t}{\pi \alpha^2 . R^2} . T_t . T_r . \exp.(-2\sigma R) \qquad \dots\dots[2]$$

Where P_t is laser source power in Watts
α is beam divergence in radians of transmitted laser beam
T_t and T_r are transmission of transmitting and receiving optics for laser
And σ is atmospheric attenuation at laser wavelength
For extended target with area $A_t > (\pi \alpha^2 R^2) / 4$

$$P_r = \frac{\rho\, P_t\, A_r}{\pi\, R^2} \cdot T_t \cdot T_r \cdot \exp(-2\sigma\, R) \qquad \dots\dots[3]$$

For cooperative target with retro reflector, having reflectivity r

$$P_r = r \cdot P_t \cdot (A_r / \pi\, \alpha^2 \cdot R^2) \cdot T_t \cdot T_r \cdot \exp(-2\sigma\, R) \qquad \dots\dots[4]$$

Of all the parameters that can be controlled in range finder design, beam divergence α plays very important in deciding maximum range capability of laser range finder. The maximum range capability also depends upon noise level of detector and background illumination of target. Generally the beam divergence of a laser range finder for ranging ground target is controlled to fill the targets at maximum range i.e. ranging armored vehicle divergence of laser is kept at 0.4 milli radians in ideal case. If $(P_r)_{min.}$ is the minimum power detector can detect, the maximum range $R_{max.}$ range finder can detect is given by

$$(P_r)_{min.} = (\rho\, P_t\, A_r / \pi\, \alpha^2\, R^2_{max.}) \cdot T_t \cdot T_r \cdot \exp(-2\sigma\, R_{max.}) \qquad \dots\dots[5]$$

The returns from the target depend on reflectivity and cosine of angle of incidence and obey Lambert's Law. The diffuse reflectivity of rough target depends on the constitution of wavelength and to a lesser extent on the angle of incidence. The reflectivity varies widely, value of a few percent is often assumed. Smooth surfaces often show a marked variation with angle of incidence. Targets have better reflectivity in visible and near infrared wavelengths. Generally for natural and extended targets it is taken as 0.4, while for military targets it is taken as 0.1 for wavelength near infrared. Target reflectance characteristics and reflection of various paints on metal and natural target in visible and infrared region are given in reference [17]. Under clear meteorological visible range of 10 km. atmospheric attenuation factor $\exp(-2\sigma\, R)$ is taken as 0.5 for lasers operating in visible or near infra red range of light.

10.4 LASER RECEIVER

The maximum range capability of a laser range finder, also very much depends on receiving lens aperture, transmission coefficient T_r of receiver, background noise, quantum efficiency and noise figure of detector. The detectors used in visible or near infrared is background noise limited. Directivity of laser receiver, generally taken is 20 to 30% less than laser beam divergence. Receiver aperture is taken to match its dark current noise with background noise. Using narrow band interference filter or tuned detector, background noise can be reduced. The peak signal to rms noise ratio in a laser receiver [18], when ranging a target in the Earth atmosphere is given by

$$\frac{\hat{S}}{N} = \frac{\beta^2\, P_r^2\, R_L\, G^2}{2eB(\beta\, P_b + I_d)\, R_L\, G^2 + 2\, k\, F\, T\, B} \qquad \dots\dots[6]$$

Where $2eB\beta P_b R_L G^2$ = Noise due to background illumination of target by Sun

$2eBI_dR_LG^2$ = Noise due to detector dark current.

$2FkTB$ = noise due to amplifying system.

$$P_b = \frac{(H_{\lambda s} + H_s\chi)\alpha^2 A_r T_r}{4} \left[\rho\, e^{-2\sigma R} + \frac{\sigma_s(1 - e^{-2\sigma R})}{\sigma}\right] \qquad \ldots\ldots\ldots\ldots[7]$$

If the noise due to detector dark current and preamplifier noise is much smaller than the back ground noise due to sunlight then,

$$\frac{\hat{S}}{N} = \frac{\beta\, P_s^2}{2e\, B\, P_b} \qquad \ldots\ldots\ldots\ldots\ldots[8]$$

For less background noise P_b, the detector should have high quantum efficiency at laser wavelength and the receiver should be highly directional with a narrow bandwidth B_0 interference filter.

The Sun irradiance H at a point outside the Earth's atmosphere is

$$H = \int_0^\infty H_\lambda\, d\lambda \qquad \ldots\ldots\ldots[9]$$

= 1345 W m^{-2} at aphelion

= 739 W m^{-2} over detector response.

The spectral irradiance at sea level is H_λ

At 6943 A^0 = 0.12 W m^{-2} A^{-1}

At 1.06 μm = 0.06 W m^{-2} A^{-1}

At 2.06 μm = 0.01 W m^{-2} A^{-1}

Therefore, while working at higher wavelength, not only the background radiation due to the Sun is less, but its contribution due to backscatter from atmospheric particles will also be less.

For pulse range finders, direct detection techniques i.e. detection of intensity and variation of the light intensity on photo detector are used. The amplifiers used have bandwidth of 15 to 40 MHz depending upon the pulse width used and lower cut off frequency is at 0.1 MHz.

For continuous wave and frequency stable lasers, heterodyne detection techniques [19] are used. In this, optical signal is mixed with a stable coherent optical local oscillator by a beam splitting mirror or other optical summing devices as shown in figure-1.

Both are then directed towards the input of a photo detector where mixing action takes place between the signal and the local oscillator field. Optical mixing has become practical because the narrow emission with the intermediate frequency in the microwave region or lower is well within the bandwidth of fast photo detectors, the photo mixing is called optical heterodyne. For this purpose, the laser frequency should be highly stable. One such radar operating at 1.06 μm is described by Kene [20].

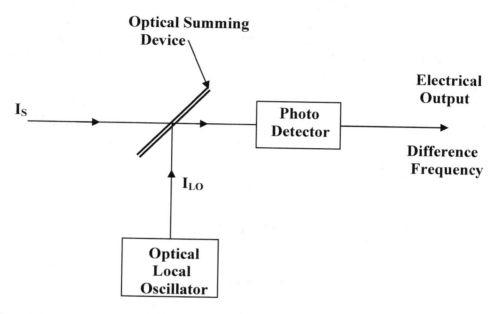

Figure-1 Optical Heterodyne Detection Techniques

10.5 DESCRIPTION

10.5.1 Solid State Ranging System

The majority of laser range finders in operation use optical pumped solid-state laser as the source of transmitted power [21,22]. The principal upon whom these systems work are common to all, the difference between various instruments lies in the fine details of the optical design and of the electronic circuitry.

The optical elements of ranging systems are shown in figure-2. The range finders have three optical channels i.e. transmitter, receiver and aiming sight. For visible and near infrared laser range finders, the receiver and sighting axes are combined through a dielectric coated beam splitter separating laser and visible light received by receiving cum objective lens. The collimating optics of laser source is a beam-expanding telescope having magnification m so that laser transmitter beam divergence α is m time less than beam divergence α_s of laser source i.e.

$$\alpha = \alpha_s / m$$

..................[10]

232

The aperture of receiving cum objective lens depends upon range requirement, generally, it is 40 to 70 mm depending upon detector used and noise due to back ground radiation. Sighting / aiming telescope magnification used for ranging ground target is generally taken as x7 having field of view 6 to 7^0. Before the detector, narrow bandwidth interference filter of 50 to 100 A^0 bandwidth is used to filter out back ground radiation. Sighting optics uses image inverting prism for compact and rugged field design with graticule on KG-3 filter glass gives eye protection from any scatter or reflected laser light, as KG-3 absorb laser i.e. near infrared radiation not visible to eye. Laser range finder with common transmitting and receiving optics using polarizing beam splitter is generally avoided as detector life is very much reduced due to back scatter of laser laser radiation from source on common optics.

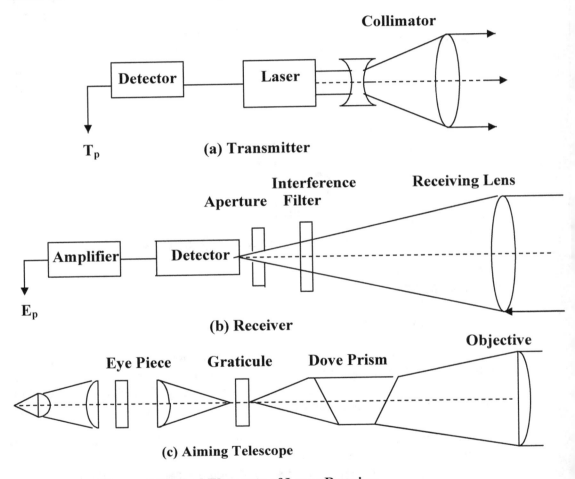

(a) Transmitter

(b) Receiver

(c) Aiming Telescope

Figure-2 Optical Elements of Laser Ranging

Figure-3 illustrate essential parts of a solid state pulse laser range finder used for ranging extended ground targets unto 20 km using low repetition rate giant pulse laser source. Laser source uses a Q-switched element- rotating prism or Electro-optic switching arrangement.

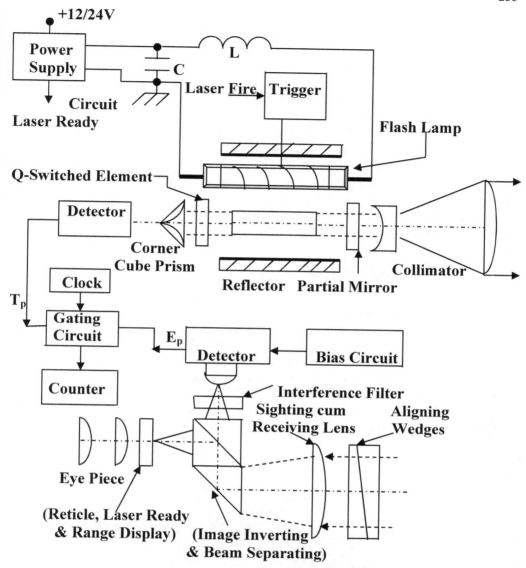

Figure-3 Block Diagram of Laser Range Finder

Most common Q-switching element for Nd:YAG laser material are passive reverse bleachable dye embedded in plastic or Cr^{+4}:YAG, this has advantage as delay / synchronizing circuit is not required. Beam expanding telescope using negative lens is used on laser axis to reduce beam divergence. Aligning wedges are used in transmitting and receiving beam path for high degree of alignment for lasers with low divergence laser beam and high directive receiver. Laser ready indication and range display are positioned from the eyes in such a way that laser aiming can be carried out conveniently without changing eye position. Laser ready and range display is projected in the eyepiece, as this eyepiece is close to the eyepiece of sighting cum aiming telescope.

Operation of Laser: With switching on the laser power supply, high voltage fly back converter charges energy storage condenser, as soon as energy storage condenser is charged to required voltage, which takes fraction of second to few seconds depending upon laser source, converter is switched off and LED in eye piece indicates laser ready. Aiming mark in sighting telescope is placed on target to be ranged, keeping the aiming mark on target; of pressing the range fire switch a giant pulse is generated in laser within few tens of microseconds to a few hundred milliseconds depending on Q-switching method used. A portion of laser energy falls on p-i-n photo detector- 1 placed in laser source giving T_p as reference pulse which opens gating circuit and allow quartz clock frequency- 15, 30 or 150 MHz depending upon range accuracy required to be counted in digital counter. Depending upon distance of target diffuse light in the form of electric pulse at detector output is amplified to a TTL value E_p closes the gate and count registered in the counter are displayed as range on digital display in eye piece. Depending upon clock frequency a range resolution of 10, 5 or 1 meter is registered, as from a target at 3 km it takes 20 μ second for light to travel to and fro from range finder to target.

10.5.2 TEA CO$_2$ Laser Ranging System

A TEA CO$_2$ ranging system developed at IRDE [23] is illustrated in figure-4. It consist of DC –DC converter operating on 24V DC supply. A sealed plasma tube as shown in figure-14 in chapter-3 is used. Laser beam is collimated by x10 beam expanding telescope to a divergence of 0.5 milli radians. A large aperture germanium lens is used to collect reflected laser echo from target is received on small area MCT cooled by liquid nitrogen. Three-digit counter using clock frequency of 15 MHz. to represent one count as 10 meters to measures time of travel of laser pulse. An aligning aid of x5 is used to align range finder with thermal sight. The laser power is 0.6 MW in

Photo-1 TEA CO$_2$ Laser Rangefinder

pulse duration of 40 nano seconds and range of 5 km achieved in dust conditions of extended targets. Photo-1 shows TEA CO_2 laser range finder developed at IRDE.

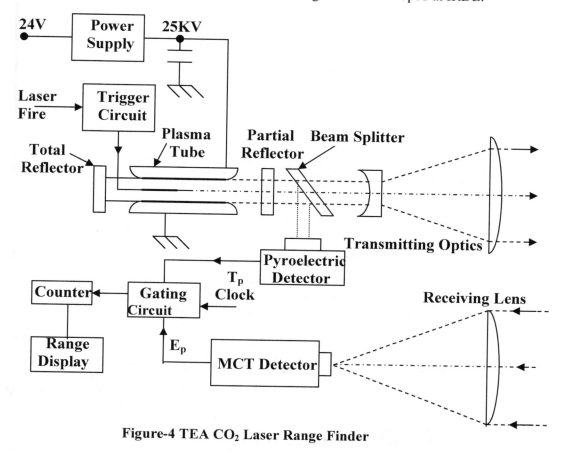

Figure-4 TEA CO_2 Laser Range Finder

10.5.3 Distance Measuring Equipment

Laser based distance measuring equipment uses intensity sinusoidal modulated He-Ne or GaAs lasers, modulated with frequency of 0.15, 1.5 and 15 MHz in steps to represent one waves as 1000, 100 and 10 meters respectively. The equipment as illustrated in figure-5 consists of laser transmitter, receiver and sighting telescope. A zero cross over detector circuit generates a pulse when sinusoidal signal crosses zero from positive to negative. Thus spacing between pulses generated between reference oscillator and echo signal received from target gives the distance of target as measured by microprocessor. With averaging techniques, the distance between laser and target can be measured within an accuracy of 5 mm. If pulse GaAs laser source is used, the principal of time measurement is given in chapter-7. The range capability of equipment depends on divergence as given in relation [5]. A He-Ne laser with beam divergence of 50µ radian can measure range of target up to few tens of kilometers.

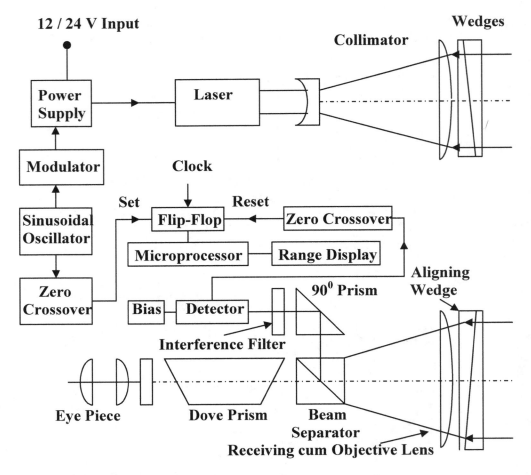

Figure-5 Schematic of Distance Measuring Equipment

10.6 LASER RANGE FINDERS

Starting with ruby laser range finder developed at IRDE in early seventies, the various types of range finders developed till now are as follows

10.6.1 Portable Laser Range Finders

Photo-2 shows a ruby laser range finder mounted on a tripod. It consists of a trans-receiver unit with detection electronics and a power supply unit with energy storage condenser bank. Rotating prism Q-switching has been used with three plate sapphire resonant reflector as partial mirror. The divergence of laser beam has been reduced to 0.5 milli-radians by using x6 beam expanding telescope. A photo multiplier tube RCA type 7265 with S-20 response has been used in laser receiver having common axis with sighting telescope of magnification x7 and field of view 7^0.

Photo-2 Ruby Laser Rangefinder (1973)

Photo-3 Nd:Glass Portable Laser Range Finder with Goniometer

Photo-3 shows Nd:Glass laser range finder with built-in power source, i.e., Ni-Cd batteries. The objective cum receiving lens of aiming sight used has aperture of 50mm. A cube beam splitter is incorporated in the sight to separate laser echo, which is focused on the detector after passing through interference filter. It can measure target range up to 10 km in clear visible conditions with accuracy of ± 10 meters. The range counter can measure two target ranges in the same line of sight with a resolution of 30 meters. Goniometer can give target bearing in 360^0 azimuth and $\pm 30^0$ in elevation with an accuracy of ± 2 minutes.

10.6.2 Laser Range Finder for Armored Vehicle

Nd:Glass laser range finders in various configurations developed to be used with the gunner's sight of an armored vehicle. Photo-4 shows the production model of laser range finder, where the laser trans-receiver unit is mounted on mantlet of tank with gunner and commanders control units with range display mounted inside the turret. The range counter has minimum blocking range variable from 400 to 4000 meters and can measure range of two targets intercepted by laser beam.

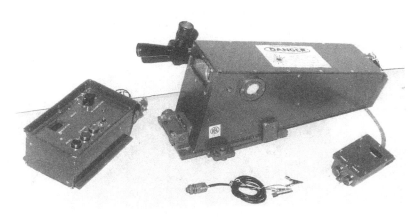

Photo-4 Nd:Glass Laser Range Finder for Armored Vehicle (1979)

10.6.3 Hand-held Laser Range Finder

Photo-5 shows the hand-held laser range finder developed at IRDE. It uses Nd:YAG laser material with passive Q-switching with minimum pulse energy of 15 mJ in a pulse duration of 8 ns. This range finder measures a range of extended target up to 10 km in clear visible conditions. This range finder has a bracket and can be coupled to night sight.

This range finder has been modified using a clock with frequency to range a target with an accuracy of ± 1 meter. Using a neutral density filter in transmitted beam, its minimum range achieved is 10 meter. The equipment is illustrated in Photo-6 with night sight for measuring range from 10 to 1000 meters during day and night with accuracy of ± 1 meter.

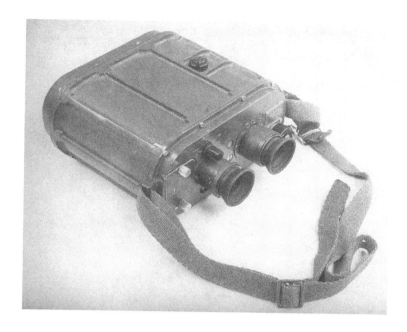

Photo-5 Nd:YAG Hand-Held Laser Range Finder

Photo-6 Nd:YAG Range Finder with Night Sight, Stand for Gap Measurement

Table – I: Specifications of Laser Range finders

Details	Portable	Armored Vehicle	Hand-held	TEA CO_2	High rep. Rate
Laser material	Ruby/Nd:Glass	Nd:Glass	Nd:YAG	CO_2	Nd:YAG
Output Energy(mJ)	70/40	40	15	25	60
Pulse-width (nsec.)	50/30	30	9	40	16
Type of Q-Switching	Rotating Prism	Rotating Prism	Passive Dye	-	Electro-Optics
Beam Divergence (mrad)	1	1	1.5	0.4	2
Detector Directivity (mrad)	2	2	3	0.4	3
Receiver Aperture (mm)	50	39	45	150	110
Maximum Range (km)	0.2/0.15	0.4-10	0.2-10	0.2-4	0.4-10
Accuracy (m)	±10	±10	±10	±10	±5
Sight (a) Magnif. (b) FOV	7^0	- -	x7 6.5^0	- -	- -
Weight (Kg)	25/8	12	3	-	11
Operating Temp.(^{0}C)	0 to 50	-20 to +50	-30 to +50	0 to +50	-20 to +50

10.6.4 High Repetition Rate Laser Range Finder

Photo-7 shows Nd:YAG laser range finder with pulse repetition rate of 10 pulse per second for ranging military aircraft targets up to 8 km range in clear visible conditions. It uses Electro optic Q-switch with transmitter energy of 60mJ, a beam divergence of 2 m.rad. This unit has aligning aid and can be coupled with radar, IR or TV tracking devices. Table – I gives the specification of various range finders developed at IRDE

Photo-7 Nd:YAG High Repetition Rate Laser Range Finder

10.6.5 Laser Ranger cum Designator

A diode pumped Nd:YAG laser range finder and designator has been recently developed at IRDE. Block diagram of source developed is illustrated in figure 14 of chapter-3 (Type of Laser Sources) Photo-8 shows this ranger cum designator. It uses electro optic Q-Switching with precise pulse coding. Specification of this range finder is given below

Specifications Laser Ranger cum Designator

Laser	Diode pumped Nd:YAG
Wavelength	1.064 µm
Peak Power	5 MW
Pulse Repetition Rate	20 Hz (6 Codes)
Detector	Silicon APD
Designator Range	5 km
Max. Range	9995 m
Range Accuracy	$\pm$ 5
Operating Temperature	-20^0 C to $+55^0$ C
Power Supply	22 – 32 VDC
Weight	8 Kg

242

Photo-8 Laser Ranger cum Designator

10.7 COLLIMATION AND ALINGING PROCEDURE

The performance of laser range finder and its maximum range capability depends on how best - (a) The laser beam is collimated (b) The laser transmitter is aligned with receiver and aiming sight (c) All the laser echo energy collected by receiving lens is focussed on small area detector, or focussed on field stop aperture, just placed before the detector. In order to carry out, collimation, alignment and focussing of received laser echo from target, following test equipment are required

(i) Laser Energy Meter
(ii) Fast Photo-detector.
(iii) Precise Parabolic Mirror - Large Aperture & long focal length (1 meter).
(iv) He-Ne Laser
(v) Laser Simulator – Diode laser emitting at Laser Source Wavelength
(vi) Black Photographic Paper
(vii) Set of Aperture
(viii) Neutral Density Filters
(ix) Optical Bench
(x) Long Stable Table

10.7.1 Collimation

In order to carry out collimation of beam from laser source, beam expanding telescope / collimator is placed axially with laser beam from source as shown in figure-6a with help of He-Ne laser. Than parabolic mirror is placed in front of source with a small. piece of black photographic paper placed at the focus of parabolic mirror as shown in figure-6b.

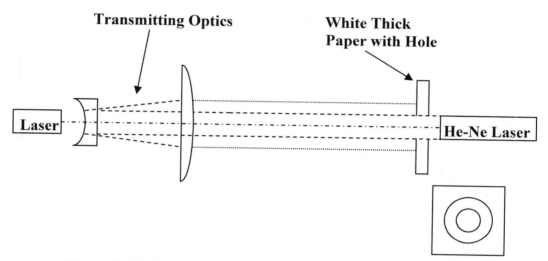

Figure-6a Centering of Laser Transmitting with Laser Source

Caution: Before operating laser, receiver aperture should be covered in order to Avoid damage to detector from laser source.

Now operate the laser and see the size of burnt spot on black paper, increase or decrease the spacing between negative and positive lens of collimator i.e. rotating the cell in which positive lens is mounted in the collimator body. Lock the moment, that is spacing between collimator lenses for minimum spot size. Now the laser is collimated, i.e. size of burnt spot divided by focal length gives laser beam divergence.

<u>Switch off the laser.</u>

10.7.2 Alignment of Laser with Aiming Sight

Now the sight is moved in its adjusting Bracket, till aiming mark is at the burnt spot of laser on black paper placed at the center of parabolic mirror as shown in figure-6b. At this position aiming sight is aligned with the laser source. The movement of adjusting brackets is locked at this point. The laser beam will fall on the target, if the laser is fired with aiming mark on the target. The accuracy of alignment can further be improved by rotating optical wedges.

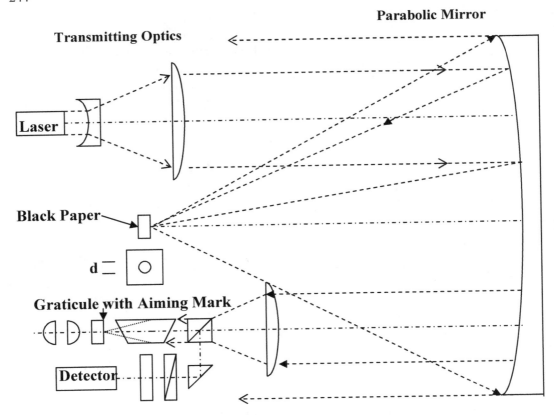

Figure-6b: Collimation of Laser Transmitter and its Aligning with Aiming Sight

Caution: While carrying out alignment of laser source with sight, laser power supply should be switched off. Further, one should be sure that laser source, parabolic mirror or black paper with burnt spot placed at focus of parabolic mirror is not disturb, while adjusting aiming sight bracket.

10.7.3 Alignment of Receiver with Laser Transmitter / Sight: While carrying out receiver alignment, a low power diode source, emitting at laser wavelength is placed at focus of parabolic mirror as shown in figure-6c. The aiming mark of sight is placed at the active area of diode laser or pinhole aperture placed close to diode laser. The aligning wedges placed in laser receiver are rotated till maximum signal as viewed on CRO connected to detector output. Again spacing between receiver lens and detector or detector with directive aperture is a adjusted for maximum signal at output of detector. At this point rotation of optical wedges and spacing between detector and receiving lens is locked. Now the laser receiver is aligned with transmitter and aiming sight.

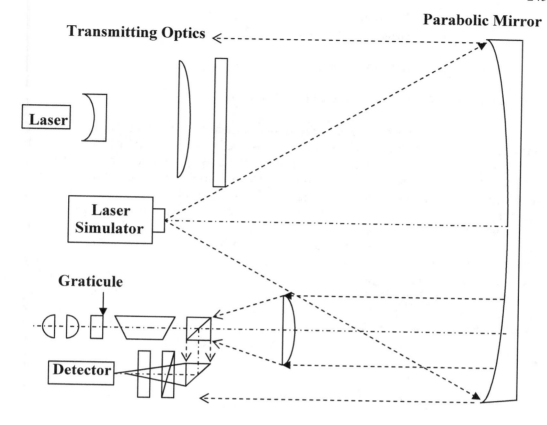

Figure-6c: Aligning of Laser Receiver with Laser Transmitter /Aiming Sight

Caution: While carrying out alignment of laser receiver with aiming sight, laser power supply to be switched off and external bias to detector being given. In case laser power supply is switched on for detector bias purpose, laser transmitter to be covered, as due to accidental fire of laser will damage diode simulator and detector of laser receiver.

10.8 RANGE FINDER OPERATION

(i)   Connect supply or place charge battery in the range finder

(ii)   Remove caps from transmitting, sight and receiver window

(iii)   Aim the graticule of sight on the target to be ranged.

(iv)   Move the blocking range control approximately less than target range.

(v)   Switch on the supply switch

(vi)   As soon as laser ready indication is displayed, press laser fire switch keeping aiming mark at target.

(vii)   Read range on digital display

10.9 TESTING OF LASER RANGE FINGER

The performance of laser range finder is measured by its aligning accuracy and maximum range capability. There are various methods followed, which are used to measure performance of laser range finder in field. Best method for its performance check is to measure its extinction ratio [24] at close target board i.e. white target board with diffuse reflection placed at distance of 500 or 1000 meters from range finder depending upon range available and maximum range capability of range finder. In carrying out this test, a set of neutral density filters are required to reduce the intensity of laser transmitted beam, till 50% ranging probability is achieved in firing laser and measuring range. Attenuation of filter value gives extinction ratio of range finder. From extinction ratio maximum range of laser range finder can be calculated for given target reflectivity under clear visible condition. The surfaces of neutral density filters should be parallel within 30 seconds of arc, otherwise aligning accuracy of laser range finder will get disturbed. If ρ_0 is the target reflectivity of target board at distance R_0 intercepting the complete laser beam, the laser power received at the detector is given by equation [3] as

$$P_{r0} = \frac{\rho_0 \, P_t \, A_r}{\pi \, R_0^2} \cdot T_t \cdot T_r \cdot \exp(-2\sigma \, R_0) \qquad \ldots\ldots\ldots[11]$$

If M fold power is reduced to a minimum detectable level by neutral density filter, then

$$(P_r)_{min} = \frac{\rho_0 \, P_t \, A_r}{\pi \, R_0^2 \, M} \cdot T_t \cdot T_r \cdot \exp(-2\sigma \, R_0) \qquad \ldots\ldots[12]$$

By comparing equation [5] and [12], we get

$$M \, \frac{\pi \, R_0^2}{\rho_0} \, \exp(2\sigma \, R_0) = \frac{\pi \, R_{max}^2}{\rho} \, \exp(2\sigma \, R_{max}) = \frac{P_t \, T_t}{(P_r)_{min} / A_r \, T_r} \qquad \ldots\ldots[13]$$

Taking logarithm to the base 10 and multiplying by 10, the so-called extinction ratio S is given by

$$S = 10 \log \left[\frac{M \, \pi \, R_0^2}{\rho_0} \, \exp(2\sigma \, R_0) \right] \qquad \ldots\ldots[14]$$

$$S = 10 \log \left[\frac{\pi R_{max}^2}{\rho} \exp.(2\sigma R_{max}) \right] \qquad \ldots\ldots..[15]$$

$$S = 10 \log \left(\frac{P_t T_t}{(P_r)_{min} / A_r T_r} \right) \qquad \ldots\ldots..[16]$$

The physical concept of the extinction ratio can be seen from the equation [16], which is a ratio of the output power P_t to the received minimum detectable power per unit area of the objective $(P_r)_{min} / A_r$. This is a comparative range sensitivity of the whole system, expressed in db.

10.9.1 Calculations

From equation [14]

$$S = 10 \log M + 10 \log (\pi R_0^2 / \rho_0) + 10 \log [\exp.(2\sigma R_0)] \qquad \ldots\ldots..[17]$$

For $R_0 = 0.5$ km, Meteorological visible range of 20 km, with neutral density filter of attenuation of N_A db. and diffuse target reflectivity ρ_o of 0.95, for 1.06 μm

$$S = N_A - 0.83 + 0.61 \qquad \ldots\ldots..[18]$$

For range requirement of 10 km for target of reflectivity of $\rho = 0.1$ and atmospheric transmission of 70% at 1.06 μm, from equation [15] we get

$$S = 10 \log (2\pi \times 10^3) + 10 \log 2 = 40.95 \text{ db} \qquad \ldots\ldots..[19]$$

From equation [18] and [19]

With neutral density filter of $N_A = 41.17$db, 50% of range obtained at target board of reflectivity 0.95, at 0.5 km, the range finder can measure maximum range of 10 km for target of reflectivity of 0.1 and atmospheric transmission of 70% over entire range, at wavelength of 1.06 μm.

10.10 MAINTENANCE PROCEDURE The laser range finder maintenance procedure is divided into three parts

Preventive Maintenance.
Routine Maintenance
Fault diagnosis, repair and adjustments

10.10.1 Preventive Maintenance is for long life and better performance of the instrument.

(i)   If the instrument is not in use for long period, it should be kept with battery removed in carrying box with dehydrated silica gel.

(ii)   Plug connectors of harness and sockets on the instruments should be kept covered with caps.

(iii)   After use, spare batteries and battery in the instrument should be kept charged, fast charging of batteries should be avoided as far as possible. Spare batteries contact points to be kept covered with electrical insulating caps.

(iv)   When instrument is not in use, transmitting and receiving window of laser ranging instrument should be kept covered in carrying case.

(iv)   If the instrument is not used for six months, it should be switched on for three to four hours, to avoid degradation of electrolytic capacitors.

10.10.2 Routine Maintenance is to avoid damage to instrument and can be used effectively in short possible time, if required.

(i)   The glass windows should be kept clean, dust to be removed with blower and frequently cleaned with A.R. grade alcohol using soft muslin cloth or tissue paper.

(ii)   Silica gel, if pink, the silica gel to be replaced by dehydrated silica gel, i.e. blue in color.

(iii)   Frequently, leakage test of instrument to be carried out and filled with dry nitrogen at atmospheric pressure.

(iv)   The bearing, adjusting screw on brackets, if used to be kept cleaned and properly greased.

10.10.3 Fault diagnosis, repair and adjustments, before repair and adjustments, following **instructions / precautions** to be followed

(i)   Study the log book and study the conditions under which instrument become faulty.

(ii)   While fixing or removing connectors, twisting of cables should be avoided.

(iii)   Instrument, if dismantled or assembled, it should be carried out on a laminar flow assembly table in a dust free room.

(iv)   During dismantling, laser, optical, reflector and Q-switched elements to be kept in dissector.

(v)   When measuring energy of laser source with energy meter, receiver lens should be kept covered.

(vi)   No voltage should be on energy storage condenser.

The laser range finder becomes defective, due to laser source becomes non-functional, i.e. no laser output, or laser receiver does detect laser echo from target, defect in laser power supply or ranging counter circuit, or misalignment of laser transceiver. The defect analysis is given as follows

Table- II a – No Laser Output (Laser Source non Functional)

Faults	checks
(a) Laser ready lamp does not glow	Low voltage supply, voltage sensing and limiting circuit.
(b) Charging of energy storage does not take	Fuse, battery, fly back inverter.
(c) Q-switched motor does not run	Range fire pulse, motor drive circuit.
(d) Motor runs continuously	+5V supply on motor drive circuit.
(e) Lamp does not fire	Lamp, delay / synchronizing circuit, trigger circuit.

Table II b – No echo pulse (Laser receiver non functional)

Faults	checks
(a) No echo pulse	Bias circuit, amplifier and preamplifier
(b) Echo pulse level very low	Receiver alignment, detector sensitivity Transmitting pulse at bias circuit / time variable gain amplifier.

Table II c – No range (Range counting circuit non functional)

Fault	check
(a) No range Display	Time interval unit, counter and range display Transmitting and echo pulse at time interval unit.

SUMMARY

Laser ranging systems are used for ranging ground targets or low flying air targets for military applications. Laser based distance measuring instruments are used to measure distance for survey purpose. He-Ne and GaAs lasers are used in distance measuring instruments, while ruby, Nd:Glass, Nd:YAG, Er:Glass and TEA CO_2 sources are used in ranging systems for military applications. Most of the range finders uses Nd:YAG material in laser range finders, due to better shelf life, efficient operation. Er:Glass is used in ranging system, as its operating frequency is safe for eye retina. TEA CO_2 Laser range finders are used in adverse battle conditions with thermal sight for viewing and aiming laser on targets, but laser range finder using CO_2 are bulky, costly with less operating and shelf life. Nd:YAG used in laser range finders have limitation for adverse weather conditions. To over come these limitations, recently components are developed to make the systems eye safe, cheap and efficient. Some of the developments taking place for recent ranging techniques are given in next chapter.

LIST OF SYMBOL USED

A_r = Laser receiver lens aperture

A_t = Target area (m^2)

B = Amplifier bandwidth

B_0 = Bandwidth of interference filter

c = Velocity of light in vacuum

e = Electron charge in Columbus

F = Noise factor of amplifier

G = Amplifier gain

H_s = Irradiance over detector response

$H_{\lambda s}$ = Spectral irradiance at laser wavelength

I_d = Detector dark current

k = Boltzman constant

M = Attenuation of Neutral density filter

m = Magnification of transmitter optics

N = rms noise level at laser receiver output

N_D = Attenuation of Neutral density filter in db

P_t = Peak laser transmitter power (W)

P_b = Background power in the laser receiver

P_s = Power received at detector due to laser echo from target (W)

P_{min} = Minimum power detector can measure

R = Range of target (m)

R_{max} = Maximum Range which range finder can measure

R_o = Known distance at which target board is placed

S = Extinction ratio

$\hat{S}$ = Peak level of signal at laser receiver output

t = time of travel of light in air

T = Absolute Temperature (^{0}K)

T_r = Transmittance of receiver optics

T_t = transmittance of Transmitter optics

α = Beam divergence of laser transmitted beam

α_r = Laser receiver directivity (radians)

α_s = Beam divergence of laser source (radians)

β = Responsivity of detector (A/W)

λ_s = Laser wavelength

χ = Transmittance of receiver optical filter outside its pass band

σ = Atmospheric attenuation coefficient (m^{-1})

σ_s = Atmospheric back scatter coefficient (m^{-1})

ρ = Target Reflectivity

ρ_0 = Reflectivity of target board

Appendix-1

Target Reflection Characteristics: Curves Representing Reflection Characteristics of various types of selected materials like metals, paints, trees/leaves, soils, and building material are given in [17]. Percentage reflectance at 1, 2, 4 and 10 micron from these curves is given in Table-1 below. Since range finders these days operate on these wavelengths where the atmospheric attenuation is minimum and efficient laser sources can be built using Nd:YAG, OPO or TEA CO_2 and its second harmonic in atmospheric window in 3 to 5 micron or 8 to 12 micron. Generally manufactures of laser range finders give capability of maximum range for topographical targets reflectivity as 0.4 and Military Targets as 0.1. While target cross section of various targets at various laser wavelength of long ranging are discussed in [19].

Table-1: Various Materials and their Percentage Reflectivity at wavelengths

Material	1 μm	2 μm	4 μm	10 μm
Aluminum, Commercially Pure	70	84	90	92
Aluminum, Weathered 20,000 hrs on DC-6 a. liner	55	70	80	75
Stainless Steel, Sand Blasted	20	32	48	60
Stainless Steel, Bare Clean	26	35	45	55
Armco, Ingot Iron, Oxidized	19	40	-	-
Dirt (Light) on Flat Black Paint	26	30	20	10
Equipment Enamel, Light Grey	-	20	30	10
Cement Aged, Building Material	30	30	-	-
Asphaltic Road Material	-	40	30	10
Coal Tar Pitch Point	-	8	8	10
Oak Leaf, Winter Color	50	-	10	10
Pine Needles	30	-	-	-
Fine Slit Loam	40	55	30	4
Grady Slit Loam, Georgia	30	45	20	4
Dublin Clay Loam	20	35	15	4
Water Surface at 90^0 Angle	35	35	35	30
Water Surface at 60^0 Angle	-	5	4	2
Calcium Carbonate	94	90	60	-
Sodium Carbonate	90	80	2	-
Polished Copper	90	90	96	96
Pure Nickel, Polished	58	76	88	95
Maury Slit Loam, Tennessee	50	52	32	3
Vereeniging, Africa, Soil	50	50	22	5
Ideal Masonry, No. 100, Sy Blue	-	65	40	11
Gloss Enamel, White	-	63	44	8
Galvanised Iron, 22 MIL, Commercial Finish	28	67	80	87

REFERENCES

1. Maiman,T.H., "Stimulated Optical Radiation in Ruby Maser", Nature, Vol. 187, pp. 493-494, August, 1960.

2. Zuev V.E. "Laser Beam in the Atmosphere", Translated from Russian by James S.Wood, Publisher Consultants Bureau, New York and London.

3. Hinkley, E.D. and Kelly,P.L., "Detection of Air Pollutant with Tunable Diode Lasers", Science, Vol.171, No. 3972, pp. 635-639, 19 Feb., 1971.

4. Weibrig,P.,and Savanberg,S., "Versatile Mobile Lidar System for Environmental Monitorind", Applied Optics (USA), Vol. 42, No.33, 20 June 2003, pp. 3583-3594.

5. Mansharamani,N. Greenhouse Effect-Lidar Techniques, ISBN 81-7525-789-X, Sita Publisher, Jakhan, Duhradun-248001, (2006)

6. Sam, R.C., "Alexandride Lasers", Handbook of Solid State Engineering, Optical Engineering, Vol.18, Publisher Marcel Dekker, Inc. USA.

7. Mansharamani, N., "Role of Photonics for Detection of Explosive", Optics and Optoelectronics – Theory, Devices and Applications, Vol.2, pp.1006-1011, Norosa Publishing House, New Delhi & London, 1999.

8. Mansharamani, N., "An Electro-Optical Warning Device for the Trains", Communications in Instrumentation, Vol. 5, No.2, pp.83-86, (1997).

9. Measure, R.M., "Laser Remote Chemical Analysis", John Wiley and Sons, New York, 1988.

10. Chaspy,P.C., Yoh-han Pao, Siult Kwong and Eugene Nodov, "Laser Optoacoustic detection of explosive vapors", Applied Optics (USA), Vol.15, No.6, June 1976, pp.1506-1509.

11. Pearson, G.N., Harris, M., Willetts, D.V., Tapster, P.R. and Robert P.J., "Differential Laser Absorption and Thermal Emission for Remote Identification of Opaque Surface Coatings",Applied Optics (USA), Vol.36, No.12, pp.2713-2719.

12. Poropat, G., "Quake Alert: Laser Ranging Gives Japanese an Early Warning", Photonis Spectra, Vol.31, No.5, pp.152-159, May 1997.

13. Benson, R.C. and Mirarchi, M.R., "The Spinning Reflector Techniques for Ruby Laser Pulse Control", IEEE Transactions on Military on Military Electronics", Vol. 8 No.1, January 1964, pp. 13-21.

14. Juyal, D.P. and Vasan, N.G., "Eye Safe Solid State Lasers for Range Finders", Proc. SPIE, Vol.3729, 1999, pp.282-286.

15. Price, W.F. and Uren, J., "Laser Surveying", Publisher Van Nostrand Reinhold, London pp.162-220, 1989.

16. Mansharamani, N., "Solid State Laser Rangefinders – A Review", Defence Science Journal (India), Vol.45, No.4, 1995, pp.315-324.

17. Jelalian, A.V., "Laser Radar Systems", pp.213-278, Publisher Artec House, Boston, London, 1992.

18. Electro Optics Handbook, RCA-EOM10, 1968.

19. Harney, R.c., "Comparison of Techniques for Long Range Laser Ranging", Proc.SPIE Vol.783, Radar II, 1987, pp.91-100.

20. Kene, T.J., Koylovsky, W.J., Beyers, R.L., and Byvik, C.E., "Coherent Laser Radar at 1.06µm using Nd:YAG Laser", Optics Letters, Vol.12, No.5, 1987, pp.239-241.

21. Forrester, P.A. and Hulmes, K.F., "Review – Laser Range Finders", Optics & Quantum Electronics, Vol.13, 1981, pp. 253-93.

22. Scheht, R.G., Paul, J.L., "Advances in Miniature Lasers for Ranging Applications", SPIE, Vol.247, Advances in Laser Engineering and Applications, 1980, pp.116-123.

23. Juyal, D.P., Barthwal, N.K., Singh, A.L., Gupta, S.P., Rudrappa, M.T., Saha, T., Sindhwal, P.S., Gupta, M.C., Raturi, D.C., Jain, M.C., and Roy, A.K. "Design and Development of a Pulsed CO_2 Laser Range Finder", J. of Optics (India), Vol. 17 No.4, October-December 1988, pp. 91-93.

24. Hu Jixian, "The Index and Testing of Extinction Ratio for Pulsed Laser Range Finder", SPIE, Vol.1230, pp.173-175.

CHAPTER-11

RECENT DEVELOPMENTS IN RANGING TECHNIQUES

11.1 INTRODUCTION

The ranging techniques involve laser sources, low level detection, and signal processing and range computation. The laser sources mainly used in laser range finder at present are ruby, Nd:Glass, Nd:YAG, Er:Glass, and TEA CO_2 operating at 0.6943, 1.06, 1.054, 1.55 and 10.6 μm. For distance measurement for survey purpose are He-Ne and GaAs operating at 0.6328 and 0.85 μm. At present most of the laser range finders are using Nd:YAG laser source and distance measuring GaAs laser source with pulse operation. The detectors used for low-level detection of laser echo are photo multipliers with S-20 photo cathode, silicon and InGaAs - avalanche photodiodes, or liquid nitrogen cooled MCT depending upon wavelength of laser used. Direct detection techniques are used due to operation simplicity. Range computation is carried out with time interval and digital counters using clock frequency of 15 to 150 MHz. to give range accuracy from 10 meters to 1 meter. Pulse correlation techniques have been used to measure range up to 2 km using high repetition rate GaAs laser with pulse peak power of 15 W [1]. Microprocessors are used in distance measuring equipment to increase distance measurement accuracy by time interval averaging techniques.

The range finders in operation have following limitations, due to adverse weather conditions target cannot be viewed, hence ranged. In dust conditions, viewing and ranging is possible with thermal sight and ranging using TEA CO_2 laser range finder, but due to cost, less life shelf life and cooling problems, moreover this range finder have maximum range limitations under humid and foggy atmosphere. Due to these reasons, TEA CO_2 laser range finders are seldom used. The Nd:YAG laser is not safe to eye. Er:Glass laser are inefficient due to three level operation and it cannot be passively Q-switched, making it bulky and complex. To overcome these drawbacks / limitations some of the devices and components under development will find application in range finders in future.

Therefore in this chapter, system and devices under development for fiber communication may be used in range finders. Viewing and ranging is possible under adverse weather conditions using range-gating techniques with gated intensifier channel tube [2]. In this chapter, the recent development in component and devices, viewing and ranging under adverse weather conditions, pulse correlation techniques for double pulse coded laser source and microwave modulated optical radar are described.

11.2 RANGING & SIGHTING WITH RANGE GATED LASER ILLUMINATION

11.2.1 Viewing & Ranging Under Adverse Weather Conditions

256

Under adverse weather conditions, like thick fog or under water detection of target or water depth measurement, it is difficult to locate or view targets, even at close ranges with passive sights including thermal sights. This is due to lot of backscatter from fog and other particles, when target is illuminated even with laser. With the help of gated micro channel plate (MCP) intensifier [2], it is possible to locate target and measure range from the delay of gated pulse to MCP intensifier tube from the instant of laser pulse transmission. The requirements of laser are that it should operate in nanosecond pulse duration with pulse repetition rate of about 30 pulses per second or less with wavelength having high quantum efficiency for the cathode material used. The multi alkali photo cathode responds to light in the 400 to 900-nm range. A Gen. II bialakli photo-cathode deposited on negative affinity material GaAs has response in the 200 to 800 nanometer (nm) ranges. The gallium arsenide Gen. III covers a 580 to 900 nm band. Indium (In) doped GaAs photo cathode extent the range of the Gen. III to a greater than 1.06 μm. The laser used are second harmonic Nd:YAG with Gen. III green cathode which has 52% quantum efficiency at 532 nm with gating interval as short as 5 ns for under water detection and depth measurement. While Alexandrine lasers operating in the range of 700 to 825 nm are used for range-gated illumination [3] with Gen. III red photo cathode intensifier tube having maximum quantum efficiency of 23% at 800 nm. The Nd:YAG laser is not used for illumination because Gen. III NIR has quantum efficiency of 0.01% at 1.06 μm.

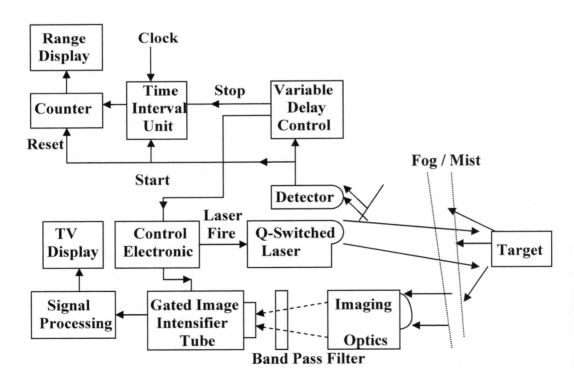

Figure-1 Range Gated Illumination and Ranging

The basic principle on which range and gated illumination is possible is illustrated in figure-1. The laser beam with some divergence is aimed towards known direction of target or beam is scanned and opening of gate of MCP intensifier tube is delayed in steps from the time of laser transmission, till the target is visible. From this delay range R of target visible can be computed as

$$R = \frac{1}{2} t_d \cdot v$$

[1]

Where v = velocity of light in medium.

t_d = delay time between gating pulse to intensifier tube and laser pulse transmission.

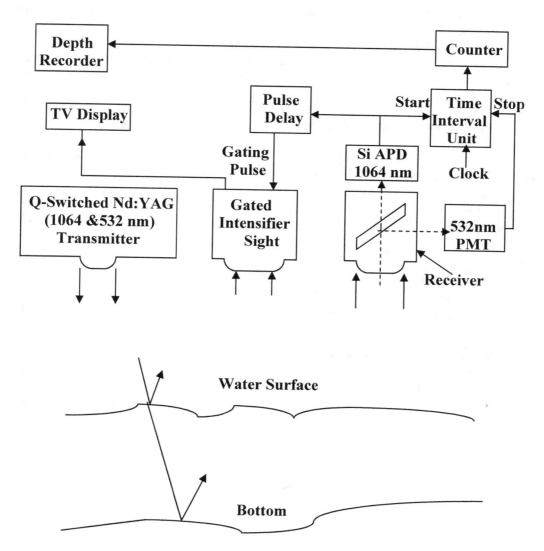

Figure-2 Under Water Illumination and Depth Measurement

258

11.2.2 Laser Bathymetry

Laser bathymetry is used to measure depth of shallow water up to depth of 30 meters. It consist of a high peak power Nd:YAG laser source [4] operating at 1.064 μm (1064 nm) along with second harmonics at 532nm with sharp pulse of 5 ns duration or less and repetition rate of 20 to 30 pulses per second. The instrument is mounted in a pod of aircraft flying over surface of water at a point where depth is measured or object submerged under water to be viewed. The instrument is illustrated in figure-2. Both the laser beams at 1064 nm and 532 nm generated in Nd:YAG laser source with second harmonic generating crystal KTP ($KTiOPO_4$) are transmitted from aircraft onto the surface of water, where 1064 nm beam is reflected from surface of water while laser at 532 nm is transmitted inside water and get reflected from land surface at bottom or any submerged object, the reflection at 532 mn is received on laser receiver fitted with photo multiplier. The bottom of water or submerged object can be viewed in TV connected to sight fitted with gated intensified video camera. The delay is adjusted with respect to signal received from surface of water. The depth can be measured using relation [1] given above.

11.3 LASER RANGE FINDER WITH CODED DOUBLE PULSES

A double pulse coded [5] Nd:YAG passive Q-switched laser has been developed at IRDE to range target under background noise / against Electro optic counter measure. The delay between pulses can be varied from 150 to 250 μs. Each pulse generated has energy of 10 mJ and pulse width of 8 ns. Flash lamp is fired twice at variable interval depending upon code, to discharge first condenser directly and another through SCR to

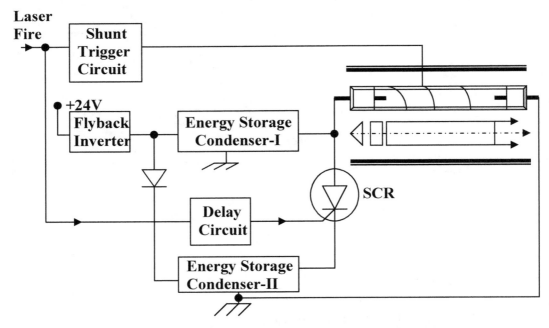

Figure-3 Passive Q-Switched Nd:YAG Laser (Double Pulse Coded)

pump laser rod twice as shown in figure-3. Using the pulse correlation circuit, noise is eliminated to get correct range of target. This type of laser range finder can be used under noisy background and for identifying friend and foe (IFF) role in military applications.

The block diagram of pulse correlation circuit is shown in figure-4, the transmitted pulse

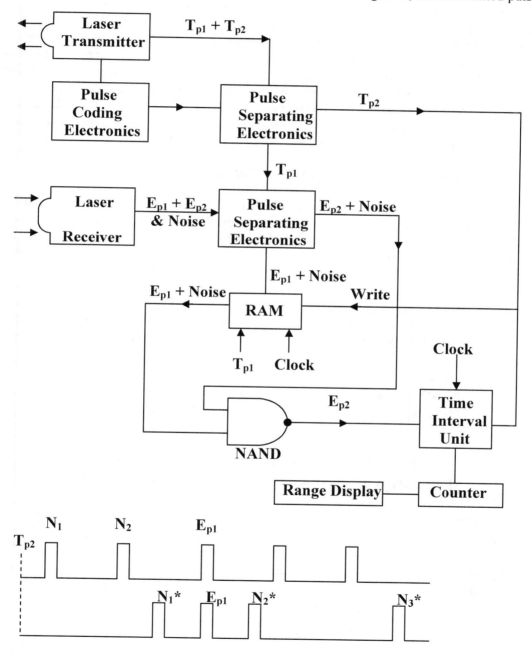

Figure-4 Pulse Co-Relation Technique (Double Pulse Coded Laser Range Finder)

generated at coded time interval are separated. The laser echo pulses from detector output of laser receiver at the instant of first laser transmission are stored in static Random Access Memory (RAM), this will consist of range echo pulse with random spaced pulses due to background shot noise, internal or external noise source. At the time of next laser pulse transmission, detector output from laser receiver corresponding to target echo and noise pulses are compared with output in RAM corresponding to target echo and noise pulses as a result of first laser transmission. The echo signal from target at the instant of laser transmission spaced in same time interval with respect to transmitted pulse and is allowed to pass through gate to stop ranging counter to indicate correct range of target as shown in figure-4. This type of laser source can be used for ranging fast moving air targets. The time interval between transmissions of laser pulses is so short, i.e. 250 μs that even target mowing at speed of 1000 meters per second will move only 0.25 meters in 250 μs. The range finder using this type of source can be used in armored vehicle and can indicate friend and foe status, as laser pulses are time coded.

11.4 MICROWAVE MODULATED OPTICAL RADAR:

The high power quasi continuous wave (QCW) laser diodes with efficiency of 40 to 50% efficiency [6] can be used for microwave modulated optical radar. The schematic of this type of range finder is illustrated in figure-5. The QCW laser diode is operated for pulse duration equal to twice the time required by light to travel and come back from target at maximum distance / maximum range of interest. If maximum range required for range finder is 15 km, pulse width of laser diode output should be 300 μs. The QCW laser diodes are commercially available with operating pulse width unto 400 μs with peak power of 4800 W depending upon number of arrays [6,7] used. For ranging aerial targets laser beam is collimated within few milli radians and intensity modulated at microwave

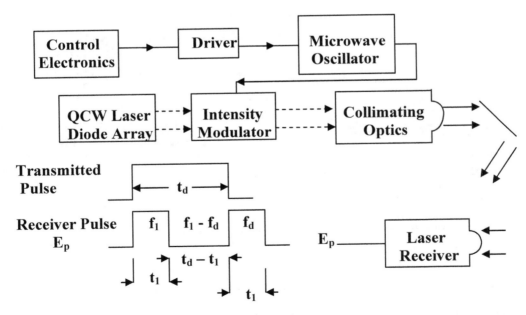

Figure-5 Microwave Modulated Pulse Optical Doppler Radar

frequency using Electro-optic modulator. The sighting is generally carried manually with optical day sight or Video camera or thermal imager. The laser echo is received by large area receiving lens on detector along with some signal from laser transmitter. The pulse echo received consists of intensity modulation with frequency f_1 for duration equal to $2R/c$, where c is velocity of light and R is target distance, $(f_1 - f_d)$, where f_d is Doppler shift in frequency due to motion of target during period (t_p-2R/c) as shown in figure – 5b and f_d for duration of $2R/c$, f_d is given by relation

$$f_d = f_1 (1 \pm v/c) \qquad\qquad ....[2]$$

Where v is velocity of target, thus measuring this duration of frequency shift and difference frequency, range R and its velocity can be computed. Thus these diodes can be used directly for Light Modulated Doppler Radar.

11.5 DEVICES & COMPONENTS OF FUTURE RANGING TECHNIQUES

The new class of device technology, high speed semiconductor and microprocessor, detectors and laser sources offers compact, reliable, less cost, better life and performance to laser ranging techniques. With these devices, it will be possible to detect, view range targets under adverse weather conditions. The following critical components and devices developed or under development will be commercially available for uses in range finders are

The Micro, fiber and quantum cascade lasers, saturable absorbers for eye safe lasers, non-line crystals for efficient harmonic generation in mid infrared spectrum of light, compact photo multipliers, ultra fast semiconductor detector with high quantum efficiency at selected wavelength of light and uncooled detectors in mid infrared spectrum of light will be some of the components and devices for use in future ranging techniques.

11.5.1 New Crystalline Host Materials for Nd: Recently New host materials like Yttrium Vanadate ($Nd:YVO_4$) and Potassium-Gadolinium Tungstate crystals $Nd:KGd(WO_4)$ /Md:KGW host materials are high efficient and better materials especially for diode pumped laser sources for range finders. Oscillation threshold is 0.1 –0.3 Joules with diode optical to optical efficiency of 60%. Their thermal conductivity and hardness is much less than Yttrium Alumanium Garnet (YAG).

11.5.2 Pump diode Lasers: The pump diode lasers GaAlAs operating at 800 nm and InGaAs operating at 965-985 nm of light are useful for pumping Nd:YAG laser material and Er and Yb doped laser glasses. These diode lasers are commercially available [6,7] operating in quasi-continuous or continuous mode at room temperature. The duration of pulse operation can be matched to lifetime of fluorescence of laser materials for efficient Q-switched operation. Their pumping efficiency and operating life is much more than flash lamp.

11.5.3 Saturable Absorber: Saturable absorber Co^{2+}: $MgAl_2O_4$ and Co^{2+}: ZnSe crystals [8,9] recently developed are useful for Q-switched operation of Nd^{3+}:$YAlO_3$ and

Er^{3+}:Yb^{3+} laser material operating at wavelength 1.34 and 1.55 μm for compact eye safe laser range finders.

11.5.4 GaN Laser: A new type of semiconductors has been developed operating in ultraviolet, blue and green spectrum of light. The GaN epitaxial layer over grown on sapphire substrate and the AlGaN /GaN modulated doped strained super lattice cladding

1. **Ni/Au Contact P-Electrode**
2. **0.05-μm Layer of P-Type GaN:Mg**
3. **0.1-μm Layer of Mg-Doped GaN**
4. **200-μm Layer of P-Type $Al_{0.2}Ga_{0.8}N$:Mg**
5. **0.1-μm Layer of Si-Doped GaN**
6. **$Al_{0.14}Ga_{0.84}N$/GaN Modulation-Doped Strained-Super lattice Cladding Layer of 120 25-μm Layers of Si-Doped GaN Separated by 25-μm Layer of Undoped $Al_{0.14}Ga_{0.84}N$:Mg**
7. **0.1-μm Layer of N-Type $In_{0.1}Ga_{0.9}N$**
8. **3-μm Layer of GaN:Si**
9. **GaN Buffer Layer**
10. **$Al_{0.14}Ga_{0.84}N$/GaN Modulation-Doped Strained-Super lattice Cladding Layer of 120 25-μm Layers of Mg-Doped GaN Separated by 25-μm Layer of Undoped $Al_{0.14}Ga_{0.84}N$:Mg**
11. **$In_{0.15}Ga_{0.85}N$/$In_{0.02}Ga_{0.98}N$ Multiple-Quantum-Wells Structure Consisting of Four 35-μm Layers of Si-Doped $In_{0.15}Ga_{0.85}N$/$In_{0.02}Ga_{0.98}N$ Quantum Well That Form a Gain Medium Separated by 150-μm Barrier Layers of Si-Doped $In_{0.02}Ga_{0.98}N$**
12. **TiO_2/SiO_2 N-Type Electrode**
13. **Sapphire Substrate**

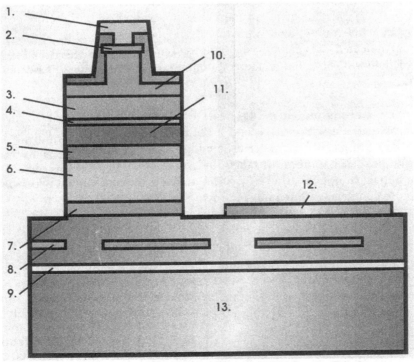

Figure-6 InGaN Multiple Quantum Well Laser Diode

layers grown on the GaN eptixial layer have life more than 10,000 hours for continuous room temperature operation [10]. This laser has quantum efficiency of 11.6% at 520 nm. A multi quantum well laser is illustrated in figure-6.

11.5.5 Q-Switched Miniature Laser: Miniature Q-Switched laser for use in compact laser rangefinder has been developed at IRDE using neodymium doped potassium-gadolinium tungstate crystal (Nd:KGW) laser rod of 3 mm diameter and 15 mm long as shown in figures-7. This laser rod is side pumped with a single quasi continuously wave (QCW) laser diode array producing 800 W at 808 nm wavelength. The laser diode array has an emitting area of 3.6 mm x 10mm. The light output from the laser diode is coupled by lens duct with facet dimensions of 2.2 mm x 12mm with pump light transmission of 77%. A highly reflective aluminum foil was wrapped around the outer half of the laser rod, which was kept in contact with the lens duct, in order to improve pumping efficiency and to dissipate waste heat. The laser resonator comprised of a 4-meter radius of curvature plano-concave high reflecting mirror and a plane output partial mirror with reflectivity of 57% at 1.06 μm. mirror spacing is about 60 mm. The pump laser diode array produced 74 mJ energy at 808 nm under drive current of 80 A in pulse duration of 100 μs. 57 mJ of light energy from lens duct was coupled to laser rod. The laser produced about 15 mj polarized normal laser output corresponding to an optical to optical efficiency of 26%. Passive Q-Switching using M/s Kodak BDN dye of 0.34 optical density produces 4-5 mJ energy in 8-9 ns pulse duration, corresponding to a peak output power of 0.5 MW. Efficiency of this laser source can be improved with antireflection coating at 808 nm on lens duct.

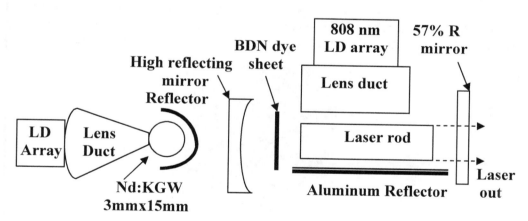

Figure-7 Q-Switched Nd:KGW Miniature Laser

11.5.6 Q- Switched Fiber Glass Laser: The high repetition rate fiber glass lasers doped [12,13] with Er^{3+} and Yb^{3+} ions have been developed using diode pump lasers for better efficiency using Acoustic Q-switching. But due to slow Q-switching, these lasers give broad pulses of few tens of nano seconds. But with the development of saturable absorbing crystals as indicated in para (b), sharp pulses with sub nano seconds duration have been produced to make these lasers useful for accurate ranging with more reliability.

11.5.7 Quantum Cascade Semiconductor Laser: The Quantum cascade semiconductor lasers using varying layers of GaInAs / AlInAs(InP) or InSb /InAlSb have been developed [14,15,16] to operate in mid-infrared spectrum of light. These lasers, for injected electron, generated 25 to 30 times the number of photons as compared to bipolar electron-hole pair devices. A 100 mW power at 130 K is reported to be generated and can be operated efficiently in pulse mode at higher temperature. These lasers can be expected, if used in laser range finder using thermal sight under adverse weather conditions.

11.5.8 Non Linear Crystals: Non linear crystals $AgGaSe_2$ have been developed to generate second harmonic in mid infra-red spectrum of light from isotope CO_2 laser operating at 9.27 μm [17]. Periodically poled lithium niobate (PPLN) is used for efficient optical parametric generation using diode pumped laser in mid infrared. These lasers can be used with MCT detector cooled by thermoelectric coolers. At mid infrared natural targets has more reflectivity. Thermal sight also does not need liquid cooling for MCT and these detectors can be cooled with thermoelectric coolers.

11.5.9 Resonant–Cavity Enhanced Photo-detector: Resonant-cavity enhanced detectors [18,19] have been recently developed to detect light in near infra-red region spectrum of light i.e. for GaAs, Nd:YAG and eye safe laser using laser pulses of sub nanoseconds duration. The quantum efficiency of conventional detector structure is governed by the absorption coefficient of the semiconductor material, meaning that thick active regions are required for high quantum efficiencies. Thick active region reduces device speeds because of the long transit time required. During past decade, a new fast device has emerged with enhanced performance and ultra fast speed by placing the active device structure inside a Fabry-Perot resonant micro cavity. This has resulted in: wavelength selectivity and a large efficiency at selected resonant wavelengths as a result

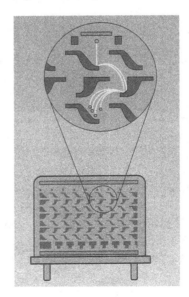

Figure-8 Compact Photo Multiplier Tube (Permission and Courtesy of Laser Focus World)

of large enhanced optical field around active region. The photo devices based on this principle are known as resonant cavity enhanced (RCE) photodiodes. These photo detectors are ultra fast, with wavelength selectivity and high quantum efficiency at desired wavelength. If such type of detectors is used in laser received, noise due to background will not be limiting factor for direct pulse detection systems. More over laser pulses of sub nanoseconds can be detected for better accuracy with less energy generated by micro lasers.

11.5.10 Compact Photo Multiplier Tube: A commercial PMT [20] based on the metal channel dynodes are recently introduced with a small TO-8 package as shown in figure-8 with a gain of million with frequency response of 300 MHz. The power supply used for this PMT draws a fraction of power from battery. This type of PMT is much cheaper as compared to channel multiplier.

11.6 CONCLUSION

As discussed in this chapter, main limitation in using Nd:YAG laser source operating at 1064 nm in laser range finger is that its range capability very much depends on atmospheric visibility. The TEA CO_2 laser range finder gives better range capability, when used with thermal sight. But due to limitation in shelf life and tedious cooling with liquid nitrogen, NdYAG laser range finders are mainly used. These laser range finders are slowly being replaced with eye safe Er:Glass range finders. But now with efficient pumping of Nd:YAG with diode pump lasers in micro size and development of ultra fast photo detector and compact photo multipliers, Nd:YAG laser source with second harmonic at 532 nm is used for ranging and gated illumination using intensified tubes with cathode, whose efficiency is 52% at 532nm. This wavelength has good transmission in water as well through thick fog. But these days high power solid-state laser sources in mid infrared wavelength are being developed. These sources uses efficient nonlinear crystal like PPLN pumped by diode pumped Nd:KWG or Nd:YVO4 high power lasers. Thermal sights at these wavelengths can be used with thermo-electric cooled photo detector MCT. These sources if used in range finders will give better range capability. Micro and fiber Q-switched laser sources will be more used with pulse correlation using fast microprocessor in compact laser range finders. The cheap and reliable GaN lasers operating in green blue region with conversion efficiency of 12%, with operating life exceeding 10,000 hrs can also be used in range gated illumination and pumping Alexandride laser, replacing second harmonic Nd:YAG lasers.

SUMMARY

The laser ranging techniques are being constantly getting updated with cheap, efficient and reliable sources with fast-high quantum efficiency detectors in receiver. With pulse correlation techniques using microprocessors, it is possible to use lasers with less operating power for better accuracy and less false alarm rate. In this chapter, new types of ranging systems are described, like range-gated illumination and ranging, pulse microwave modulated optical radar. Recent development in devices and component like reliable and efficient GaN semiconductor lasers, diode pumped Nd:YAG micro lasers and fiber lasers and ultra fast resonant photodiodes are described, these component will be used in future range finders to make them more compact, efficient with infinite life. Laser sources operating in mid infrared are under development (OPO) using nonlinear crystal like PPLN pumped by diode pumped Nd:KWG or Nd:YVO$_4$ and these sources can be used with thermoelectric cooled MCT detector for ranging with thermal sight under poor visible conditions.

REFERENCES

1. Vukas, B., "New Family of Miniaturized Laser devices", SPIE issue on Signal and Image Processing System Performance Evaluation, vol. 1310, 1990, pp. 215-221.

2. Sturz, R.A., "Laser Illuminated Range Gated Viedo", Product Literature, M/S Xybion Electronic Systems, San Diego, CA, USA.

3. Sam, R.C., "Alexandrite Lasers", Handbook of Solid State Lasers, Ed. Peter K. Choe, Vol.18, pp.444-449, Optical Engineering, Pub. Marcel Dekker Inc, USA, (1989).

4. Gaba, S.P., "Underwater Ranging", Defence Science Journal (India), Vol.34, No.1, pp. 71-78, (Jan.1984).

5. Mansharamani, N., Prasad G.R., Vasan, N.S., Sheel, T.K., and Rawat, G.S., "Double Pulse Operation of Passive Q-Switched Nd:YAG Laser", Journal of Optics (India), Vol.23, No. 2, pp.83-86, (1994).

6. Product Catalogue, M/S SDL, Inc., San Jose, CA, (USA) (1996/97).

7. Product Catalogue, M/S Thomoson-CSF, ORSAY Cadex, France (1996).

8. Kon Stantin, Yumashev, V., "Saturable Absorber $Co^{2+}:MgAl_2O_4$ Crystal for Q-Switch of 1.34 μm $Nd^{3+}:YalO_3$, and 1.54 μm $Er^{3+}:Glass$ Laser", Applied Optics (USA), Vol. 38, No.3, 20 October 1999, pp. 6343-6346.

9. "$Co^{2+}:ZnSe$ Saturable Absorber Q-Switch for the 1.54 μm, $Er^{3+}:Yb^{3+}$ Glass Laser", Eds. Pollock, C.R. and Bosembury, W.C., Advanced Solid State Lasers, Vol.10 of OSA Trends in Photonic Series, pp.148-151, Pub. Optical Society of America, Washington DC, (1997).

10. Nakamura, S., "Blue Lasers Meet Tough Commercial Requirements", Photonics Spectra, May 1998, pp. 130-135.

11. Zayhowski, J., "Q-Switched Microchip Laser Find Real-World Applications", Laser Focus World, Vol.35, No.8, August 1999, pp.129-136.

12. Mitchard, G. and Waarts, R., "Double-Clad Fiber Lasers to Handle high Power", Laser Focus World, Vol.35, No.1, January 1999, pp.113-115.

13. Minelly, J.D., "Fiber Lasers", Photonics Spectra, June 1996, pp.128-136.

14. Sirlori, S., "GaAs / $Al_xGa_{1-x}As$ Quantum Cascade Laser", Appl.Phys. Lett. (USA), Vol.73. No.24, 14 December 1998, pp3486-3488.

15. Meyer, J.R., et al, "High Temperature Mid-IR Type Quantum Well Lasers", Proc.SPIE, Vol.3001, February 1997, pp.308-309.

16. Hardin, R.W., "Diode Lasers", Photonics Spectra, April 1998, pp.110-114.

17. Schunemann, P., "Non-Linear Crystals Provides High Power for the mid-IR", Laser Focus World, Vol.34, No.4, April 1998, pp.85-90.

18. Unlu, M.S., "Resonant-Cavity enhanced devices improve efficiency", Optoelectronis World, Detectors, Supplement to Laser Focus World, March 1998, pp.15-20.

19. Davison,A. and Marshland, R., "Demand for High Speed Detectors Drive Research", Laser Focus World, Vol. 34 No. 4, April 1998, pp.101-108.

20. Winer, R.L., "Photo multiplier Advances Create New Market", Laser Focus World, Vol. 34, No. 6, June 1998, pp.107-114.

CHAPTER-12

LASER SAFETY

12.1 INTRODUCTION

The basic hazards due to lasers are

(a) Damage to the eyes – burns of the cornea or retina or both.

(b) Damage to the respiratory system from hazardous particulate and gaseous materials.

(c) Electrical hazards.

(d) Cryogenic hazards.

(e) Skin burns.

(f) Chemical hazards.

The hazards due to laser sources required for range finding, as discussed in chapter-3 of this monograph are only (a), (c) and (d). Therefore in this chapter we shall discuss only these hazards. Starting from structure of eye and its function, protective standards, laser beam hazard classification and evolution, safe range calculation based on maximum permissible exposure (MEP), preventive measures, precautions, first aid and do's and don'ts are given in details.

12.2 STRUCTURE AND FUNCTION OF EYE

The basic structure of eye is shown in figure-1. It basically consists of lens, which forms image on retina. As soon as image is formed on retina, with some minimum value of intensity of light illumination, electrical signals generated, gets transmitted to visual center of brain through optical nerve, where comparison with already stored signal takes place to recognize the object. Most sensitive part of retina is fovea. The fovea is the region of most distinct vision in daylight. The ability of eye to see details, distinguish colors and perceive depth is most highly developed in the fovea. When we speak of 20/20 vision, we are measuring visual activity in the fovea with a set of norms. The reminder of the retina is involved principally with the peripheral vision and visual activity decrease away from the fovea. Burns in the peripheral retina if small may not be noticeable by the victim, whereas, damage to the fovea, which is only about 1mm in diameter, can result in serious impairment of vision.

The power of lens changes depending upon object distance by cliary muscles to focus the image of object on retina i.e. known as adaptation. Cornea is transparent coat of the human eye, which covers the iris and the lens. It is the main refracting element of the eye.

The circular pigmented membrane called iris, lies behind the cornea of the eye. The variable aperture in the iris through light travels towards the interior region of the eye is known as pupil. The pupil dilates when exposed to low level of illumination and constricts when exposed to bright fields. The variation in pupil size extends over a wide range of lighting intensities. When viewing a bright object in sun illumination, the pupil diameter may be as small as 2 mm. Under very dark conditions, pupil dilates to 8mm. to receive more light on retina from object in dark. The aqueous and vitreous fluids as shown between iris to lens and lens to retina, transmit light between 310 to 1300 nanometers (nm). These fluids absorb light below 310 nm and above 1300 nm. Therefore retina is safe for radiation below 400 nm and beyond 1300 nm.

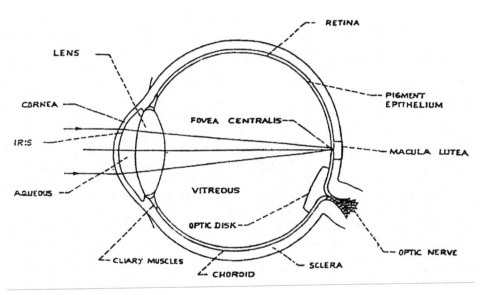

Figure-1 Basic Structure of Eye

12.2.1 Interaction of Electromagnetic Radiation with Eye

- X-rays / γ–rays shorter than 0.01 nm pass completely through eye with relatively little interaction with eye.

- The x-rays and uv-radiation between 0.01 to 310 nm are absorbed at cornea of the eye.

- Light between 310 to 1300 nm are transmitted is absorbed by retina after it is focused by eye lens.

- The longer wavelength of the mid and far infrared i.e. 1.3 to 100 μm is absorbed at the cornea or in the intraocular media before reaching the retina.

- Microwave and radio frequency radiation pass through the eye, but certain bands of wavelength, particularly 1.0 and 10 cm. wavelength are absorbed strongly in the lens and constitute a major cataract hazard.

- There are two transition zones at 300 nm and 1300 nm that affect cornea, retina, lens and iris, can cause overall damage to light vision.

12.2.2 Cause of Damage

Most radiation that falls upon the pigmentary layers of the iris and the retina whether infrared, visible or longer ultraviolet, is absorbed and converted into heat. The absorption of heat result in the rising of temperature, that leads to the destruction of tissues due to coagulation of proteins. The posterior part of the eye, particularly, the retina, is however more vulnerable because the eye, due to its focusing action, concentrate the incident energy onto a small area and raises energy level on the retina resulting in a localized lesion.

12.3 ENERGY OF LASERS FOR RANGING

As discussed in the chapter-3, laser sources used for ranging and distance measuring application and their energy /power level, pulse width, pulse duration and divergence for various roles is given in the table-I.

TABLE-I Laser sources for ranging and distance measuring equipment
A-Pulsed

Sources	Energy mJ	PulseWidth Ns	Divergence m rad.	Pulse Rate pps	Use
Nd:YAG (532 nm)	20 - 50	5 –10	0.5 –1.0	20	Ranging & Illumination Underwater
Ruby	20-100	20-50	0.5-1.0	Single	Ground to Ground Targets
GaAlAs	0.1-0.2	20-50	0.5	1000	Distance Measurement
Nd:Glass	50-500	20-50	0.5-3.0	0.1-10	Ground to Ground, Air to Ground & Vice Versa
Nd:YAG	5-200	5-20	0.5-2.5	0.1-20	Ground to Ground, Air to ground & Vice Versa
Er:Glass	20-50	20-50	0.5-1.0	Single	Ground to Ground
TEA CO_2	50-100	50-100	0.5-2.5	0.1-20	Ground to Ground, Air to Ground & Vice Versa

B-Continuous

Source	Power	Divergence m rad	Use
He-Ne	1 mW	0.1	Survey- Distance Measurement
GaAlAs	5 mW	0.5	Survey- Distance Measurement
CO_2 Frequency Stable	10 – 20 W	0.5 – 2.0	Ground to Ground, Ground to Air

12.4 LASER BEAM STANDARDS & SAFETY GUIDELINES

In laser beam, the energy distribution in TEM_{00} mode is approximately Gaussian in shape. The aperture size is such that 90% of laser energy can pass, i.e. 1/e intensity points as shown in figure-2.

Hot spots are defined as area of variance in the energy density or power density of the beam, where this density is much greater than the average across the beam. Hot spots are of considerable concern because it has been observed that under some conditions, a hot spot may develop energy /power density 10^4 times higher than the average across the beam.

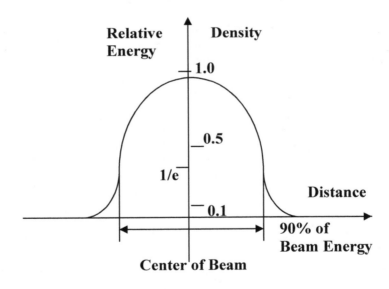

Figure-2 The Laser Beam Profile for TEM_{00} Mode Showing 1/e Intensity Point

There are two types of **safety guidelines**

(a) British – BS-4803 (1983)

(c) **American safety guidelines i.e.** by Bureau of Radiological Health (BRH) and the American National Standard Institution (ANSI).

(i)   **BRH** – Set of instruction for the manufacturer of laser products. The BRH is that section of the Federal Drug Administration in the US Department of Health, Education and Welfare, which is responsible for enforcing product performance standards. The BRH produces a standard entitled "Performance Standard for Laser Products" (1985).

(ii)   **ANSI** - Set of standard for users of these products.

The "ANSI" produces a standard entitled, "Performance standard for Laser Products", the latest version of which was published in 1985. The ANSI and the BS-4803 standards are very similar, but not identical. The main difference being that whereas the BS standards invisible lasers are classified in class 1 or class 3B. In ANSI standards invisible lasers can be categorized in class-1 and in class-3A and class-3B and has also class-2A category. In BS4803 class-2A category is same as class-1.

Labels attached to all products in both the standards contain all the information in a single label rather than on several individual labels. The word "CAUTION" is included on product labels for Class-2 and Class-3A lasers and "DANGER" must be included on those for class -3B and class-4 lasers. Examples of warning labels are shown in figure-3.

12.5 LASER CLASSIFICATION & SYSTEM REQUIREMENTS

CLASS - I   SAFE OPERATION

Operating power and energy levels requirements

For continuous wave (cw) Lasers- Power in micro watts (μW)

Equal or less than
0.4 - Nd:YAG (Second Harmonic) & He-Ne
80 – GaAlAs
100 – GaAs
200 – Nd:YAG
800 – CO_2

For Pulsed, Q-Switched Lasers – Energy in micro Joules (μJ)

Equal or less than
0.2 – Nd:YAG (Second Harmonic)
0. 4 – Ruby
2 – Nd:YAG
8 – Er:Glass
80 – CO_2

PRECAUTIONS – *No specific*

CLASS – II   **CAUTION** IN OPERATION

   *Blink reflex gives sufficient protection

Operating Power Requirements in milli Watts (mW)

For cw- lasers

1 mW - Nd:YAG (Second Harmonic) laser , He-Ne laser

LABEL -   *HAZARD WARNING (Figure-3a)
        *LASER PRODUCT – in black on yellow background.
        *DO NOT STARE INTO THE BEAM – in white on red
        background.

   PRECAUTIONS : *Reflecting Surfaces in the vicinity should be removed.
              *The beam should not be aimed at personnel and should be
               aimed above below eye level.
              *To be viewed on diffuse surfaces.

CLASS - IIIA   **CAUTION** IN OPERATION.

   Operating Power and Energy requirements

5 mW or less for cw visible lasers and 25 W per meter square maximum
  irradiance

   For pulsed Q- Switched Lasers, Energy in milli Joules (mJ)

Equal or less than
74 mJ cm^{-2} for Nd:YAG (Second Harmonic) Laser.
310 mJ cm^{-2} for Ruby Laser.

  LABEL   *HAZARD WARNING LEVEL (Figure-3b)
        *LASER RADIATION, CLASS 3A LASER PRODUCT, in black on
         yellow background.
        *DO NOT STARE INTO THE BEAM, in white on red background.
        *OBTAIN SAFETY OFFICER'S APPROVAL FOR USE OF OPTICAL
         INSTRUMENT - in white on blue background.

 PRECAUTIONS: *In addition to those required for class-2.
            *A Laser Safety Officer (LSO) should be appointed in this class
             of laser is to be used.
            *Only trained and qualified personnel should operate these
             lasers.
            *Accidental viewing of the laser by means of a telescope or other

optical instruments must be prevented – Optical viewing should be undertaken after approval from LSO.
*A key-operated switch (like that of a car ignition key) must be incorporated in order to secure from operation by untrained personnel.
*The beam must not be unintentionally misdirected by mirrors or lens.
*Warning sign should indicate the location of the whole beam, and a physical boundary to the laser hazard area.

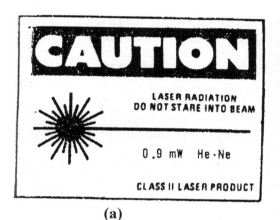

(a)

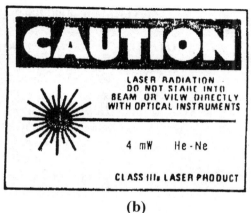

(b)

(c)

Figure-3 Labels on Laser Instruments Showing (a) Class-II, (b) Class-IIIa, (c) Class-IIIb

CLASS-IIIB   DANGER IN OPERATION

Operating Power and Energy Level Requirements

For cw- lasers output power equal or less than 500 mW, beam irradiance not to exceed 25 Wm^{-2}

For pulsed lasers: Energy not to exceed
0.16 J cm.$^{-2}$ - Nd:YAG Laser.
10 J cm.$^{-2}$ - Er:Glass Laser.
10 J cm.$^{-2}$ - CO_2 Laser.

LABEL - *DANGER WARNING LEBEL (Figure-3c)
*LASER RADIATION, CLASS –3B, LASER PRODUCT** in black on
yellow background.
*AVOID EXPOSURE TO THE BEAM,** in black on yellow background.
*A label indicating the position of laser aperture is required.

PRECAUTIONS: *In addition to those required Class-2 and Class -3A lasers
*Looking directly into the beam by naked is hazardous.
*Class-3B laser should be mounted rigidly to prevent accidental
movement.
*Deliberate optical viewing with unprotected eyes must avoided.
*All operators working in the laser hazard area must wear suitable
safety eye wears and protective cloths may be required.
*Routine ophthalmic examination is necessary for those workers
using Class-3B laser range finders.*

CLASS-IV   DANGER IN OPERATION

POWER > 0.5 W

Note: Not generally used for laser ranging applications.

12.6 MAXIMUM PERMISSIBLE EXPOSURE (MPE) & HAZARDOUS RANGE

MPE is the maximum laser power / energy density, if falls directly on person does not affect his eye i.e. retina, cornea, iris, eye lens or skin. The value of laser power varies with type of laser and the injury it produces. The energy of laser sources used in ranging is such that it affects retina, if a person is directly looking directly in laser beam with full iris aperture i.e. 7 mm. Due to finite beam divergence, the energy density of laser beam reduces and at certain range it acquire value equal to MPE. Below this range, laser is hazardous and safe beyond this range. The value of MPE for various laser sources in the visible and near infrared used for ranging and distance measuring equipment is given below.

*The MPE value for continuos wave (cw) He-Ne laser in milliwatts per centimeter square (mW cm.$^{-2}$) for various time exposure to human eye is 1.0 mW for 10 seconds (s), 1.8 mW for 1.0s and 2.5 mW for 0.25s.

*The MPE value for continuos wave Nd:YAG (2nd Harmonic) is 1 $\mu W\ cm^{-2}$ for exposure more than 10,000 seconds time.

*The MPE value for cw GaAs laser operating at wavelength 905 nm is 0.25 mW cm^{-2} for exposure time of more than 100 seconds.

*The MPE value for cw Nd:YAG laser operating at wavelength of 1064 nm is 0.5 mW cm^{-2} for exposure time of more than 100 seconds.

*The MPE value for normal pulse ruby laser operating at wavelength of 694.3 nm for single pulse of 1 millisecond (ms) pulse duration is 10 μJ cm^{-2} and for Q-switched single pulse of duration 5 – 50 nano seconds (ns) is 0.5 μJ cm^{-2}.

*The MPE value for normal pulse Nd:YAG laser operating at wavelength of 1064 nm with pulse duration of 1 ms is 50 μJ cm^{-2} and for Q-switched single pulse of 5 – 100 ns is 5 μJ cm^{-2} and for low pulse rate of 20 pps is 0.5 μJ cm^{-2}.

*The MPE value for single pulse GaAs laser operating at 905 nm is 12.5 μJ cm^{-2}, for pulse rate of 100 pps is 1.25 μJ cm^{-2} and for 1000 pps or more is 0.75 μJ cm^{-2}.

*The MPE value for single pulse GaAlAs laser operating at 850 nm is 10 μJ cm^{-2}, for pulse rate of 100 pps is 1.0 μJ cm^{-2} and for 1 KHz. or more is 0.6 μJ cm^{-2}.

The laser Hazardous range is defined as the range of laser range finder at which laser is injurious to eye if it falls on the eye of person looking directly / indirectly in the direction of range finder. This range depends on energy of laser pulse, type of laser used and beam divergence.

12.6.1 The Hazardous Range for Direct Viewing (Figure-4a)

If E_m is MPE of laser source used in laser range finder, having beam divergence of φ radians, if R_z is maximum hazard range, beyond that range laser is safe to eye for direct looking, than R_z must satisfy the relation,

$$E_m = 4 E_s / \pi (d + R_z \phi)^2 \qquad \dots\dots\dots[1]$$

Where d is initial diameter of laser beam in meter
E_s in energy of laser source.

From equation [1], we obtain, R_z in meters, neglecting atmospheric attenuation as

$$R_z = [2 (E_s / \pi E_m)^{1/2} - d] / \phi \qquad \dots\dots\dots[2]$$

278

Example: The hazardous ranges R_z, neglecting atmospheric attenuation, for

    (a) Hand held range finder using Nd:YAG laser (1064 nm), with pulse energy 10 mJ in Q-switched operation with pulse width of 5 –10 ns and beam divergence of 1 milli radians. With MPE of 0.05 J m^{-2} for single pulse operation, calculated using relation [2] comes out to be 530 meters.

    (b) Nd:YAG laser range finder with pulse repetition rate of 20 pps, with pulse energy of 0.1J per pulse and has beam divergence of 2.5 milli radians for air target ranging. With MPE of 0.005 J m^{-2} for pulse rate of 20 pps, comes out to be 1200 meters.

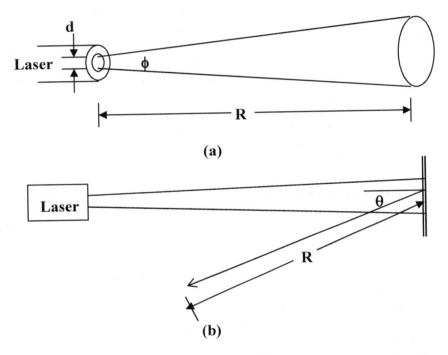

Figure-4 Hazardous Range for (a) Direct Laser Beam Viewing (b) Indirect Spot Viewing on White Diffuse Reflecting Surface

Therefore, in order to reduce hazardous range, while testing laser range finder, a neutral density filter with 2% transmission is used, where large range measurement is not possible due to crowded area for testing purpose.

KG-3 colored filter glass is used, as graticule plate in the aiming sight of laser range finder to protect eyes of operator. In case sight cannot be modified, large plate KG-3 filter can be used in front of objective of beam aiming telescope. Transmission of KG-3 filter glass is shown in figure-5.

12.6.2 The Hazardous Range for Viewing Laser Spot on Diffuse Surface:

If ρ is reflectivity of diffuse surface as shown in figure-4b, then hazardous range in viewing laser spot at maximum hazardous range R_{zD} in a direction θ for source with energy E_s is given by relation

$$E_R = MPE = \rho \ E_s \ \cos\theta \ / \ \pi \ R_{zD}^2 \qquad \qquad ....[3]$$

$$R_{zD} = (\ \rho \ E_s \ \cos\theta \ / \ \pi \ E_m \ )^{1/2} \qquad \qquad ....[4]$$

Where E_m is maximum permissible energy exposure to eye in Joules / meter square.

Example: The hazardous range for viewing ruby laser spot with energy 0.1J on white diffuse surface at an angle of 30^0 from above relation comes to a value of 2.4 metes.

NOTE: Therefore direct viewing of laser spot on diffuse surface is injurious to eye if viewed at close range in laboratory. It is advisable that laser laboratory should be well illuminated, so that aperture of iris of eye should be as minimum as possible from safety point of view.

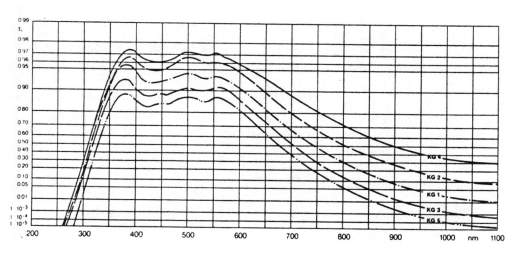

Figure-5: Transmission Curve of KG-3 Color Filter Glass with Wavelength.

12.7 CRYOGENIC HAZARDS

Cryogenic coolant such as liquid nitrogen is used for cooling MCT detector in receiver of TEA CO_2 laser range finder.

Care must be taken in handling liquid nitrogen, as liquid nitrogen can condense oxygen out of air. The liquid oxygen can condense hydrocarbons and can combine to form shock sensitive explosive mixture. Hydrocarbons must not be permitted to cool to the point where liquid oxygen will condense on them or placed where liquid oxygen may drop on them.

12.7.1 General Safety Guidelines

The hazards associated with handling cryogenic liquids are

(a). Contact of coolant with the skin will produce a **burn** and great care must be exercised to avoid contact. Therefore, asbestos gloves shall be worn when filling nitrogen pouring from Dewar and vacuum bottles.

(b). Glass vacuum bottles and Dewar shall be protected by an outer container that will not permit flying glass to escape in case vacuum bottle explode.

(c). Container shall always permit gas to escape as the cryogenic liquid evaporates.

12.8 ELECTRICAL HAZARDS

The electrical accidents occur more frequently with laser systems than the eye or skin injuries from laser radiation.

The He-Ne laser plasma tube operates at voltage between 2.5 KV to 4 KV with current of 2.5 to 10 mA depending upon laser power required. Nd:Glass , Er:Glass or Nd:YAG laser sources require energy storage condenser from 50 to 10 μf with maximum charging voltage from 1 to 1.5 KV for xenon flash lamp depending upon efficiency and power requirement for laser source in range finder. TEA CO_2 laser needs voltage of 25 to 30 KV on condenser of 10 nf. However operating personnel are well protected with proper shields, while maintenance personnel are exposed to electric hazards in course of their duties.

12.8.1 Factors Influencing Severity of Electric Shock

The degree of injury caused by electric shock is influenced by the following factors

    (i) Voltage.
    (ii) Current level through body.
    (iii) Duration of shock.
    (iv) Body resistance.
    (v) Frequency of current.
    (vi) Current pathway through body.

When current passes through vital organs such as the heart, lungs or brain, the likelihood of serious injury increase greatly with increasing current. A current pathway from the hand to the feet, as is the case for indirect accident involves the lungs and heart is more serious than shock from foot to foot. ANSI consider level over 42.5 V as hazardous, if the current exceeds 0.5 mA.

12.8.2 Physiological Effects

Due to electric shock, following Physiological Effects can occur in human body

(i) Burns.
(ii) Muscular Contraction.
(iii) Blood pressure increase.
(iv) Breathing problem.
(v) Blood and blood circulation.
(vi) Cardiac cycle failure.
(vii) Unconsciousness.
(viii) Hemorrhages.
(ix) Nervous system damage.
(x) Increase in body temperature.

12.8.3 Safety Guidelines

*All electrical cables, terminals, switches shall be properly protected and maintained to prevent accidental contact with energizing circuit.

*Capacitors must be discharged, before carrying any repair work, since capacitors can retain charges, even when the power is disconnected, bleeder resistor shall be provided across capacitor, so that capacitor remains discharged during off conditions.

*Cover over high voltage circuit shall be interlocked to prevent access to energizing components and the high voltage warning sign shall be attached to each cover.

*Proper grounding to be provided for all non-current carrying conductor, chassis, shields etc.

*Safety devices, rubber gloves and insulating mats must be provided during repair of laser power supply.

* Avoid rings, metallic wristwatch belts during repair of power supply.

* Use only one hand during working on circuit or control devices.

*Do not handle electric equipment, when hands, feet or body is wet, or perspiring or when standing on wet floor.

*Cover floor with dry matting, when working with high voltage, as floor are conducting.

*Energy storage condenser should be suitably shielded with metal to protect persons if an explosion occurs.

*The equipment operating in excess of 15 KV should be checked for possible emission of radiation.

*Water leaks should be checked and prevented from cooling system before operation.

*Flash lamp or laser plasma tube should be mounted in such a way that their terminals cannot make any contact which can result in a shock or fire hazard in a shock or fire hazard in the event of lamp or tube failure.

*Suitable warning devices to be provided.

*First Aid instructions from electric shock must be displayed at proper place.

12.8.4 First Aid Instructions from Electric Shocks:

I. SWITCH OFF. If this is not possible, PROTECT YOURSELF with dry insulting material and pull the victim clear of the conductor. DO NOT TOUCH THE VICTIM WITH YOUR BARE HANDS until he is clear of the conductor, but DO NOT WASTE TIME. Proceed as follow

(a) Lay the patient on his back. Quickly Loosen waistband and clothing round neck. Clear the throat. Wipe out any foreign matter in his mouth with your finger or cloth wrapped around your fingers.

(b) Place victim on his back. Place on a firm surface such as the floor or ground, NOT on a bed or sofa.

(c) Lift the head and tilt the head backward by putting one hand underneath the neck and the other on the crown of the head.

(d) Hold the head titled as far back as far back possible and lift up the jaw firmly, closing the lips. This keeps the victim's airway clear by straightening the breathing passage. (this will automatically keep the tongue out of the airway).

(e) Take a deep breath. Open your mouth wide and place it tightly over the victim's mouth. At the same time pinch the victim nostrils shut or close the nostrils with your cheek. Or close the victim's mouth and place your mouth over his nose. This latter method is preferable with babies and small children. Below into the victims mouth or nose with a smooth steady action until the victim's chest is seen to rise. (Inspiration)

(f) Remove your mouth to let him breath out; his chest will fall. (Expiration).

(g) Remove mouth. Listen for the return of air that indicates air exchange.

(h) Repeat. Continue with relatively shallow breath, appropriate for victim's size, at the rate of one breath each five seconds. The movement of victim's chest provides visual confirmation of the success of your efforts.

NOTE: If you are not getting air exchange, quickly recheck position of head, turn victim on his side and give several sharp blows between the shoulders blades to free foreign jam matter. After four or five breaths, stop and determine, if HEART is BEATING by checking the pulse. If the heart is beating, return to mouth-to-mouth respiration until breathing start or a physician tells you to stop.

DO NOT GIVE LIQUID UNTIL VICTIM BECOMES CONSCIOUS.

III. If the heart has stopped begin cardiac / heart massage.

(a) Lay the victim on his back on the ground or some other firm surface.

(b) Place the heel of one hand, with the other on top of it, on lower part of the   sternum (breast bone).

(c) Apply firm pressure vertically downward aided by weight of the body, about 60 times a minute.

(d) At the end of each pressure stroke, the hands are to be lifted slightly to allow full recoil of victim's chest.

(e) Sufficient pressure should be used to depress the sternum an inch or so towards the vertebral column (spine).

(a) At the end of each pressure stroke, the hands are to be lifted slightly to allow full recoil of victim's chest.

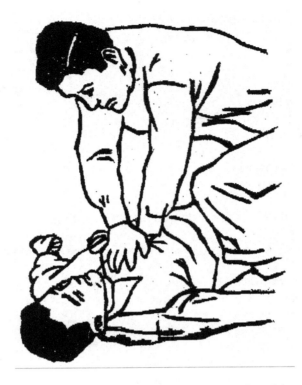

IV. Artificial respiration must continue simultaneously with external heart massage at the rate of about 5 compressions of the heart to one inflation of the lungs.

Massage should continue until the victim's pulse is clearly felt and the color returns to normal, or until medical assistance arrives.

NOTES:

*Do not attempt heart / cardiac massage if there is obvious damage to the victim's chest wall.

*There is a real danger of damage to internal organs by the improper use of external cardiac massage.

*Particular care must be taken with infants and small children, with whom much less pressure is required to depress the sternum than in the case of adults. In these cases the fingers should be used in preference to the palms of the hands.

12.9 DO'S AND DON'TS

*Laser beam should not be aimed at people, particularly their eyes.

*Avoid looking into the primary laser beam.

*Avoid looking at visible laser beam reflections, as these too have the potential for causing retinal burns.

*Avoid looking at an exposed pump source, i.e. flash lamp or gas discharges.

*Use safety eye wears to filter out the specific laser frequency, whenever risk exceeds in a place where radiation intensity is expected to be more than MPE.

*Laser should not be unattended i.e. possibility exist that unauthorized and untrained person may use it.

*Object with specular reflection should not be put in path of laser beam.

*Remove rings, watches, and shinning jewelry extra, as hazardous reflections are possible.

*Laser should be operated in well-controlled area; unnecessary people or visitor may not be allowed in that area.

*Particular caution should be exhibited in working with lasers that operate at invisible spectral region.

*Directing the laser beam at, or tracking non-target vehicles or aircraft is strictly forbidden.

*Any exposure to eye may be immediately reported to doctor, preferably ophthalmologist experiences with retinal burns.

*First aid instruction, to be displayed at proper place.

12.10 LASER SAFETY OFFICER'S RESPONSIBILITIES:

A. Establishes and administered laser radiation safety program.

B. Ensure that trained and qualified people should be allowed to use laser equipment.

C. Present during repair and testing of high power laser.

D. Not allow unnecessary presence of persons, or visitor people in hazardous area.

E. See that operator wear well-protected devices like eye wears glows, clothing.

F. Advice in case of accident or emergency.

G. Display of warning signs and first aid instructions at proper place.

H. Inspection and maintenance of laser hazardous area.

I. Inventory control and records

J. Personnel records of authorized or potential laser users

K. Management of suspected or actual laser accidents

L. Annual inventory of all lasers in any laser installation and all mobile lasers

M. Transfer, receipt and disposal of any registered laser.

N. For scheduling medical examination

O. For assuming that scheduled medical examination has been completed

P. Watch for suspected or real laser accident.

SUMMARY

Persons operating laser range finder should be conversant with laser safety aspects, especially the lasers in visible and near infrared can cause injury to retina of eye, In this chapter, starting with basic structure of eye, laser classifications depending upon power / energy is given. The maximum permissible energy (MPE) for various laser used in range finders are given. Based on MPE, hazardous zones can be calculated based on relation given. Electrical hazards are also given, where many accidents have taken place, especially to research workers. One should know first aids from electric shocks and these instructions be properly displayed in research laboratory and laser maintenance sections. General safety guidelines for eye and prevention of injury from cryogenic are also given. In the end do's and don'ts in using lasers with responsibilities of laser safety officer are also given.

REFERENCES

1. Rampal,V.V. and Prasad,G.R., "Monograph Determination of Safety Factor for Protection of Eye against Laser Radiation", J. of Institution of telecom. Engrs. (India), Vol.16, No.5, 1970, pp.325-329.

2. Gupta,A.K., "Laser Safety and Control" Lecture Notes from Continuing Education Program (CEP), A course on Laser Instrumentation organized by Mansharamani,N. from Sept. 23-27, 1996 at IRDE, Dehradun.

3. Mallow,A. and Chabot.L., "Laser Safety Handbook", Publisher: Van Nostrand Reinhold Company, New York, (1978).

4. "Laser, Safe Operation and Basic principles", a booklet from M/S Convergent Energy, Sturidge, USA and Cambridge, U.K.

5. American National Standard Institution, "Safe Use of Lasers", ANSI Standards, Z-136.1, New York (1973).

6. Price,W.F. and Uren,J., "Laser Surveying", Publisher, Van Nostrand Reinhold (International), Cambridge, U.K. (1989).

INDEX

CPSIA information can be obtained
at www.ICGtesting.com
Printed in the USA
BVHW011106300719
554669BV00006B/357/P

9 781419 698736